Piano-Hinged Dissections

Piano-Hinged Dissections

Time to Fold!

Greg N. Frederickson

CRC Press
Taylor & Francis Group
Boca Raton London New York

CRC Press is an imprint of the
Taylor & Francis Group, an **informa** business
AN A K PETERS BOOK

to Karen, who led the way, and
to Diane, who shares that great smile

Contents

Preface

This book has been a surprise for me, much more so than either of my previous books. Before embarking on this one, I had no inkling that the subject matter was even remotely possible. By comparison, the last book, *Hinged Dissections*, would seem like a foregone conclusion. When I began that one, at least I knew of a few dozen examples of swing-hinged dissections. Even then, that book seemed wildly audacious.

At first it was really difficult figuring out how to design piano-hinged dissections. Getting the pieces to land in the right place after several cuts and folds was decidedly tricky. Even after the first dozen or so successes, I wasn't sure that I could identify enough interesting material to fill this book. The progress in terms of the number of dissections was slow, sometimes frustratingly so. Yet, one insight at a time, I built up my stockpile of useful techniques, exciting dissections, and intriguing properties. As I gained confidence, I was amazed by the mind-boggling transformations made possible by the simple operation of folding.

I was stunned by the beauty of many of the dissections. And the best of the bunch still produce a sense of wonder in me. How could I resist the elegant saddle-cyclic hinging of the mitre to the gnomon in Figure 2.12? I had to love the general technique that dissected an ellipse to a heart in Figure 4.2. The transition in Figure 6.7 from an R-flap to a flap-step kept me marching in lockstep! I could hardly believe the nifty reduction method that dissected a rectangle with a hole to a square as in Figure 7.7. The way that flexible "house" collapsed in Figure 8.6 was seismic! The symmetrical placement of the three cap-cycles of the hexagons in Figure 9.32 made me feel triumphant. Fascinated, I still kept a respectful distance as I stalked the furry beast in Figure 10.10. How could I not be at a loss for words when I saw those three interlinked cap-cycles in Figure 12.6?

Once these and other discoveries set the book in motion, the next challenge materialized: to produce perspective views of the piano-hinged assemblages that would naturally convey their symmetry and complexity. After several false starts, I identified a three-step procedure: draw the as-

semblages in OpenGeometry running under Visual C++, export them to
POVray where I ray-traced and labeled them, and finally, export bitmaps
to the Unix program xv, which I used to convert them into encapsulated
postscript.

A perspective view makes it easy to enjoy the surgical precision of the
dissection of squares in Figure 13.6. And who can stay rational as we
take in the magic of Figure 14.6? Double joy awaits us with two pairs
of Greek crosses in Figures 15.4 and 15.6. The folded-out five-pointed
star in Figure 16.14, with its gorgeous inside-out property, is simply out
of this world. Aren't we triply blessed with the cap-cyclicly-hinged three
hexagrams in Figure 17.8? The two dodecagons in Figure 17.16 are enough
to make us kick up our feet in celebration!

However, even those perspective diagrams weren't always enough to
convey the excitement. So, I borrowed a videocamera, shot video clips, and
used Microsoft's Moviemaker to produce the videos for the accompanying
CD. My foray into moviemaking gave new life to the expressions "home
movies" and "hands-on." It was a one-man operation set up on our dining
room table, with the camera pointing down onto a canvas pillow snatched
from our patio furniture. I was not only the actor (supplying the hands),
but also the director, the scriptwriter, the producer, the cinematographer,
the cameraman, the art director, the editor, and the narrator. I did it all
and had lots of fun.

As you explore this enchanting new topic with the help of the figures
and videos, you may well conclude that piano-hinged dissections are challenging to discover and difficult to illustrate. Yet, paradoxically, you need
relatively little background to understand and enjoy them. As with my earlier books, the intended audience is anyone who has had a course in high
school geometry and thought that regular hexagons were rather pretty. You
should find it easy to make passable models of many of the piano-hinged
dissections, if you cut them out of, for example, a manila folder and use a
normal adhesive tape.

In addition to presenting the piano-hinged dissections, I couldn't resist
sharing the latest news about (general) geometric dissections. I was lucky
to have played a key role in locating what had been a "lost" manuscript
on geometric dissections. The author of the manuscript, Ernest Irving
Freese, was an architect in Los Angeles who, near the end of his life, be-
came passionate about dissections. Besides cataloging known dissections,
he invented many new ones, some rather brilliant, and a few that have
served as the basis for some of my piano-hinged dissections. I'm pleased to

give you a sneak preview of his work in a five-part series spread throughout this book.

You may also enjoy the short sections that I have written on related folding topics. Charmed by the standard calculus problem of cutting and folding a rectangular sheet to form a box of maximum volume, I wrote an article on both its history and a new, improved solution. The article, now included in this book, won the Polya Award for expository writing from the Mathematical Association of America. I have also crafted two additional short sections, one on folding blocks of stamps and the other on folding maps and square twists.

All in all, creating this book has been a remarkable experience for me. I certainly hope that you will have fun with it too.

I am maintaining a webpage for the book where I will post new developments and items of interest. The URL is http://www.cs.purdue.edu/homes/gnf/book3.html. I would like to hear of any new or improved dissections. My email address is gnf@cs.purdue.edu.

Acknowledgments

It is impossible to forget the many people who helped me during the researching and writing of this book. I would like to thank Walt Hoppe, who cut the pieces for some of the dissections out of flat wood panels on his laser-cutting machine. Having substantial, accurate models that I could enjoy manipulating helped to sustain my enthusiasm as I proceeded with this project. I owe an enormous debt of gratitude to Vanessa Kibbe, who plowed through a mountain of her uncle Ernest Freese's records that had accumulated over the years of his life and then lay untouched for forty-five more years after his death. Words can scarcely capture the exhilaration of at last viewing the manuscript that I had given up any hope of locating. I thank her for granting me permission to reproduce here, in electronically enhanced form, a few of the plates from the manuscript. I would like to thank William A. Freese and Dixie B. Barry, both now deceased, for sharing memories of their father with me.

I am indebted to Gavin Theobald for sharing with me his superb dissections of many-sided polygons to a square. With his generous permission I have reproduced two of his dissections, along with figures that explain how he derived them. I would like to thank Judy Lindgren for surveying a mountain of her father's correspondence for me. As in earlier books, I thank Will Shortz, for supplying citations to earlier versions of Sam Loyd's puzzles, and David Singmaster, for sharing with me an electronic copy of his work on sources of recreational mathematics and also copies of Dudeney's puzzle columns in the *Weekly Dispatch*. I thank Ivan Moscovich for sending me copies of his loop puzzle, and René Gerritsen for sending me lovely reproductions of sheets of triangular stamps.

Once again, friends old and new helped me improve the text. I thank Joseph Malkevitch for helpful comments on an early draft. I would like to thank Don Albers for suggesting that I create a video that could accompany this book. I thank Joop van der Vaart, for a helpful suggestion about the figures, and William Waterhouse, for reminding me that the angles of a triangle sum to a straight angle only in Euclidean geometry. I would like to thank Helena Verrill for assistance in identifying origami references. I would like to thank Margherita Barile, James Buddenhagen, Sheila Curl, Woody Dudley, James A. Landau, Kenneth Manders, Elena Marchisotto, John Mason, Jeremy Ottenstein, Dick Stanley, and Steve Viktora for assistance in tracking down references. Special thanks go to Woody Dudley, who made numerous suggestions that improved my original version of Folderol 1.

Many people helped me produce the graphics images in the book. Thanks to Voicu Popescu, for his help in identifying problems with some of the graphics software that I was using; to Georg Glaeser, for providing me with version 2.0 of Open Geometry in advance of its general release; and to Dan Trinkle, for assistance in converting images from one electronic format to another. I would like to thank the Persistence of Vision Development Team for making their software publicly available.

I am pleased to acknowledge the assistance of Apollonia Steele and Jim Burton at the University of Calgary library, Evelyn Blaes at the library of American University in Washington, DC, Peggy Kidwell at the Smithsonian Institution, and Ronald Roache at the Library of Congress. I would like to thank the library staff at Purdue who handled my multitude of interlibrary loan requests, as well as the libraries who generously filled those requests. I would like to acknowledge the libraries at Brown University, which I visited. I would also like to acknowledge the National Science Foundation, which supported the early part of this work through grant CCR-9731758.

I gratefully acknowledge the permission of the Mathematical Association of America to reprint my article, "A New Wrinkle on an Old Folding Problem," from *The College Mathematics Journal* (2003). I also acknowledge the kind permission of Springer Science and Business Media to include the figures and text from my article, "Piano-Hinged Dissections: Now Let's Fold!," in *Lecture Notes in Computer Science*, vol. 2866 (2003), pages 159–171, copyrighted by Springer Verlag Berlin Heidelberg 2003. I thank Cambridge University Press for permission to reproduce Figures 1.1, 3.6, 22.7, and 22.9 from my second book, *Hinged Dissections: Swinging & Twisting* (2002).

Supplementary Resources Disclaimer

Additional resources were previously made available for this title on CD. However, as CD has become a less accessible format, all resources have been moved to a more convenient online download option.

You can find these resources available here: https://www.routledge.com/9780367446253

Please note: Where this title mentions the associated disc, please use the downloadable resources instead.

Chapter 1

Time to Fold

A geometric dissection is a cutting of a geometric figure into pieces that we can rearrange to form another figure. When applied to two-dimensional figures, they are striking demonstrations of the equivalence of area that have intrigued people through the ages. The earliest examples date back to the Greek scholar Plato more than two millennia ago and to several Arabic-Islamic mathematicians a millennium ago. In the last two centuries, these curious constructions have attracted increasing attention not only from amateur but also from serious mathematicians.

Some dissections have a remarkable property, based on connecting the pieces with hinges in just the right way. We can then form one figure by swinging the pieces one way on the hinges, and the other figure when we swing the pieces the other way. It is also possible to pull the same sort of trick with hinges that allow a twisting motion. Either way, the effect is so hypnotic that, while under its spell, I wrote an entire book on the subject. Now it's time for a different kind of hinge—a piano hinge—that allows two flat pieces to fold together or apart. (The term *piano hinge* comes from the long hinge that attaches the lid of a grand piano to the box containing the strings.)

The idea of using piano hinges with "two-dimensional" dissections came to me while I was standing in the shower on the morning of Halloween, 2000. Even though my book on hinged dissections was loaded with surprising new dissections and was almost complete, I was feeling discouraged. I had realized how difficult it is to craft good models of hinged dissections. Swing-hinged dissections (to be described in Chapter 3) are subject to torques that can pull real hinges loose, and twist-hinged dissections (also in Chapter 3) rely on hinges that are not ready-made. Too bad I couldn't use piano hinges, which are stable and easily obtainable.

1

Then again, perhaps I could use piano hinges, if each polygon were a shallow prism separated into two levels which we would dissect further into flat pieces. A piano-hinge could connect two flat pieces that are side-by-side on the same level and force one piece to flip on top of the other. It didn't take long to get excited, as I realized that this is just the right insight.

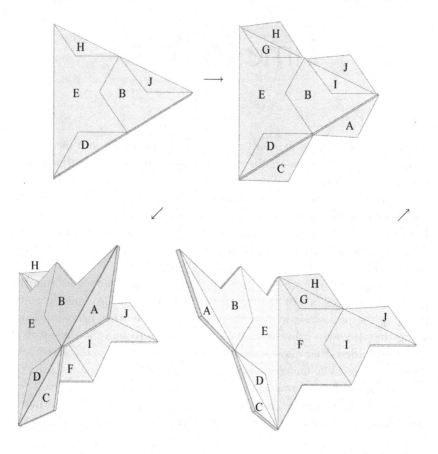

Figure 1.1. Start folding: from an equilateral triangle . . .

Furthermore, this new hinging motion is what you get when you fold a piece of paper. So it's easy to experiment with and play with this new type of hinged dissections. People who enjoy origami, the Japanese art of paper folding, may also enjoy piano-hinged dissections. Note, however, that origami is really rather different, in that it usually starts with a single square sheet of paper, doesn't involve cutting, often produces an abstract version of representational art, doesn't generally involve precise geometrical measurements, and allows folding the paper into an arbitrary number of levels, not just precisely two.

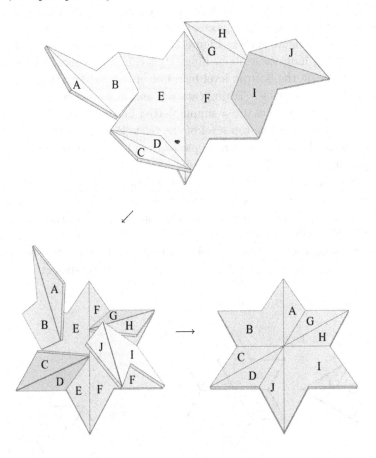

Figure 1.2. Finish folding: ... to a hexagram.

When I stepped out of the shower on that Halloween morning, I wasn't sure that I could piano-hinge any dissections at all. My goal was to design a dissection in which all of the pieces would be attached together with piano hinges into a single assemblage, so that when we folded the pieces one way on the hinges, they would form one figure (filled out on both levels), and when we folded them another way, they formed the other figure (again filled out on both levels). I quickly focused on an old standby, the dissection of a hexagram to an equilateral triangle. After several hours, I found the

10-piece dissection that I have animated in Figures 1.1 and 1.2. I was off and running! (And not to the showers!)

While the sequence of snapshots in Figures 1.1 and 1.2 helps to illustrate the folding motion, it does not do a good job of communicating the precise shape and hinging of the pieces. Instead, let's specify a piano-hinged dissection with a figure such as Figure 1.3, which explicitly describes both levels. We will indicate a piano hinge that connects a piece on the top level to a piece on the bottom level by a line of dots next to the hinge line on each of the two levels. To indicate a piano hinge between two pieces on the same level, we will use a simple dotted line segment. As we see in Figure 1.3, piece E on the top level of the triangle hinges with piece F on the bottom level. In the hexagram, pieces E and F still share a hinge, but both now sit side by side on the bottom level.

The hexagram-triangle dissection has a simple property, namely, that in either target figure, each piece resides on just one level. For example, in the triangle, piece E resides only on the top level, and in the hexagram, piece E resides only on the bottom level. We shall refer to such dissections as *folding* dissections. Moreover, since the pieces fold out flat to give a nonoverlapping net (Figure 1.4), we can call it a *nettable* dissection.

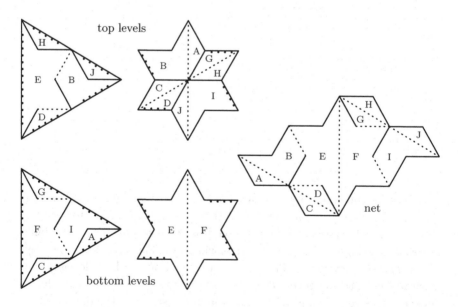

Figure 1.3. Folding a hexagram to a triangle.   Figure 1.4. Net for folding.

Can we gain an advantage by allowing pieces to extend into both levels? A year after I discovered my first piano-hinged dissection, I found a way to reduce the number of pieces for that dissection. In Figure 1.5, pieces E and I extend into both levels. Indeed, as a result of being merged with piece I, piece J in Figure 1.3 has disappeared. Continuing this sort of shell game, piece E has gained a second level as piece F has lost an equivalent amount. The shaded regions on the right in Figure 1.5 indicate the additional level added to pieces E and I. Take a look at the piano-hinged assemblage of pieces in Figure 1.6. Later on we will see more examples of this type of swapping.

We can characterize the hexagram and triangle pair as an old standby, because a 6-piece unhingeable dissection of this pair appeared in the anonymous Persian manuscript, *Interlocks of Similar or Complementary Figures*, from approximately 1300 C.E. Of several 5-piece dissections found by Geoffrey Mott-Smith (1946), one of them is the basis for the 10-piece folding dissection in Figure 1.3. Each level of the triangle corresponds to the dissected triangle as Mott-Smith gave it. Harry Lindgren (1964b) also gave this same 5-piece unhingeable dissection.

My dissection of a hexagram to a triangle is not the earliest piano-hinged dissection of figures that each have two levels. Figure 11.2 presents

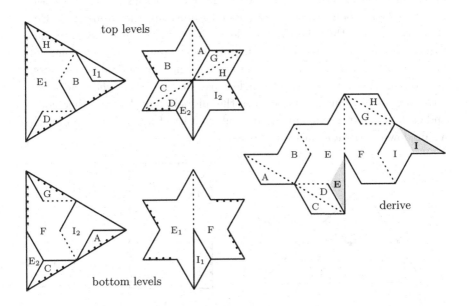

Figure 1.5.   Improved piano-hinged dissection of a hexagram to a triangle.

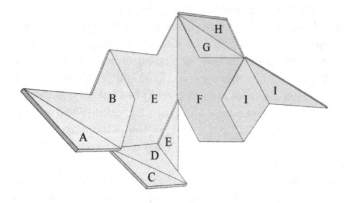

Figure 1.6.   View of the assemblage for the improved hexagram to a triangle.

a variation on the earliest example, described independently in two different variations, one by Kenneth V. Stevens (1994) and one by Jan Essebaggers and Ivan Moscovich (1994). There are also a few other antecedents that are related to piano-hinged dissections. A mathematics lecturer at Kumbakonam College in Madras, India, B. Hanumanta Rau (1888) described how to fold a triangular piece of paper to illustrate the property of Euclidean geometry that the three angles of a triangle sum to 180°. Hanumanta Rau accomplished this by folding the triangle into a rectangle of half the length and half the height of the triangle, but of double thickness, as shown in Figure 1.7. The triangle must sit on a base whose adjacent angles are acute. One fold is between the midpoints of the two other sides, and each of the two remaining folds is along a line from a midpoint of such a side down perpendicularly to the base.

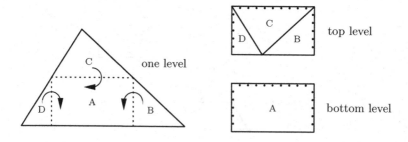

Figure 1.7.   Folding a one-level-thick triangle to a two-level-thick rectangle.

Hanumanta Rau's construction is not a piano-hinged dissection, because the triangle has only one level. However, it is easy to produce a related dissection in which each target figure has two levels. Just make a second copy of the one-level assemblage on the left in Figure 1.7, and use the original copy to fill out the top level of a triangle and the second copy to fill out the bottom level of the triangle. The result, in Figure 1.8, is a folding dissection of one triangle to two rectangles. Let's have primed letters indicate pieces from the second copy. In Figure 1.9 we are just beginning to fold together each level of the triangle to produce the corresponding rectangles with two levels.

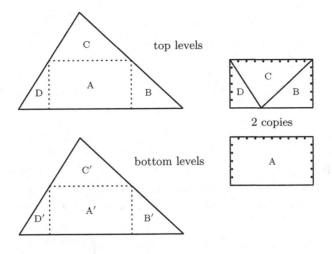

Figure 1.8.   Folding dissection of a triangle to two rectangles.

Can we find a piano-hinged dissection of the triangle into one rectangle, rather than two? The rectangle would be the same height as the triangle, and the length would be half the length of the triangle's base. We need only modify the previous dissection, piano-hinging the two levels of the triangle along its base. Figure 1.11 displays a folding dissection of a triangle to a rectangle of the same height. In Figure 1.10 we have folded down the top level of the triangle and are just beginning to fold the assemblage into a single rectangle with pieces E and A on the bottom level.

Donald Bruyr (1963) transformed a regular hexagon of single thickness to an equilateral triangle of double thickness, as shown in Figure 1.12. Just as with Hanumanta Rau's construction, Bruyr's construction is not a piano-hinged dissection, because the hexagon has only one level.

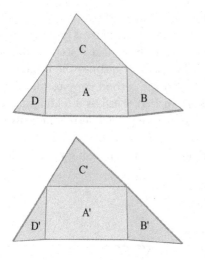

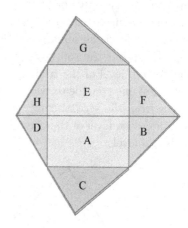

Figure 1.9. Triangle to two rectangles.   Figure 1.10. Triangle to a rectangle.

Again, it is easy to produce a related dissection in which the target figures each have two levels. Make a second copy of the one-level assemblage on the left in Figure 1.12, and use the first copy to fill out the top level of the hexagon and flip over the second copy to fill out the bottom level of the hexagon. Figure 1.13 displays the resulting folding dissection of a hexagon

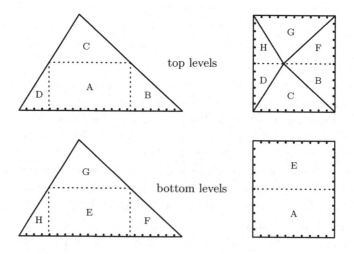

Figure 1.11. Folding dissection of a triangle to a rectangle.

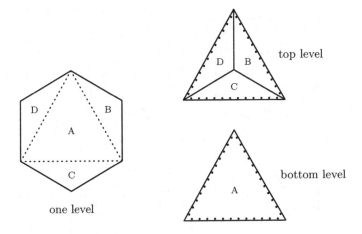

Figure 1.12. Folding a one-level-thick hexagon to a two-level-thick triangle.

to two triangles. As before, we'll use primed letters to indicate pieces from the second copy. We see the two assemblages in perspective in Figure 1.14. There is another way to piano-hinge the pieces in Figure 1.13 that makes a simple first puzzle.

Puzzle 1.1. *Find another way to piano-hinge the pieces in Figure 1.13 to give a folding dissection of a hexagon to two triangles.*

Actually, we can convert any convex equilateral parhexagon to two appropriate triangles. (A *parhexagon* is a hexagon in which opposite sides are parallel.) Just draw a triangle whose vertices are at every second vertex of the parhexagon, and then fold along the triangle's edges. The other three vertices of the hexagon will meet at the triangle's circumcenter (the point at which the perpendicular bisectors of the triangle's sides meet). We see the corresponding piano-hinged dissection in Figure 1.15.

It is disappointing that, in contrast to the history of normal geometric dissections, there is almost no history for piano-hinged dissections. Yet, these puzzles and their corresponding demonstration pieces are so nifty that I have explored related subjects that people have studied, such as forming an open box from a rectangular sheet, folding sheets of stamps, and folding maps. Short discussions of these topics will appear under the heading "Folderol." (Check out this term in a dictionary!) They provide an interesting counterpoint to the main theme of the book.

I have also crafted a multipart series about the dissections of Ernest Irving Freese, a Los Angeles architect who died in 1957. The resulting

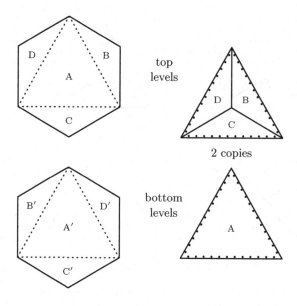

Figure 1.13.   Folding dissection of a hexagon to two triangles.

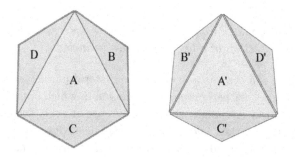

Figure 1.14.   Perspective view of assemblages for a hexagon to two triangles.

installments, starting with "The Lost Manuscript of Ernest Irving Freese," rely on materials that lay forgotten for decades in storage in his Los Angeles house and were presumed lost until recovered recently. Buried along with an amazing amount of accumulated clutter, they might easily have been discarded by descendants who wouldn't appreciate their value.

Even though Freese's manuscript contains no piano-hinged dissections, it's well worth taking a peek at here. First, Freese discovered unhinged dissections that serve as the basis for my piano-hinged dissections of seven

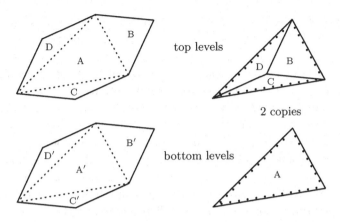

Figure 1.15. Equilateral parhexagon to two appropriate triangles.

triangles to three, three hexagons to one, and three hexagrams to one. For example, compare Freese's Plate 86 (on page 26 of this book) with my Figure 17.7. And second, these materials are so fascinating that I just couldn't resist including a few of them. I hope that these installments will help bring this extraordinary man and his beautiful dissections well-deserved, if overdue, recognition.

As in my previous books, I have included a number of puzzles for readers to solve, with solutions in the last chapter. Enjoy them, and don't fold too quickly—unless it's with paper models to test your dissections!

Finally, a few words about the organization of this book: Chapter 2 elaborates on the piano-hinged model and surveys nifty properties of piano-hinged dissections. Chapters 3 and 4 discuss other types of hinged dissections and how to convert them to be piano-hinged. Chapters 5 and 6 handle quadrilaterals and rectangles, adapting well-known techniques such as the strip, step, and P-slide techniques. Chapter 7 shows how to reduce certain piano-hinged dissection puzzles to simpler puzzles. Chapters 8 and 9 present tessellation and other techniques to handle puzzles involving a few noncongruent squares, and similarly for triangles and for hexagons. Chapter 10 gives tessellation methods for a number of congruent squares to one or more congruent squares, and similarly for triangles and hexagons. Chapters 11 and 12 explore piano-hinged dissections for pentominoes. Chapters 13 and 14 encompass dissections of squares and triangles for integer identities. Chapters 15 and 16 investigate properties of crosses and stars that yield special piano-hinged dissections. Chapter 17 examines many-to-one dissections of crosses, hexagrams, and dodecagons. Chapters 18 and 19 concern mixed sets of figures, related by some particular trigonometric relationship.

However, you need not follow the sequence of chapters carefully. I planned the book to be one that you could dip into just about anywhere and still enjoy this lovely new recreation. Indeed, I have selected fifty-some dissections to demonstrate in videos that you will find on a compact disc in the pouch on the inside back cover of the book. Watch a video and then read the corresponding discussion in the text. Or identify the bold letter **(W)** for wooden model or **(C)** for card stock model in a figure caption, and then go watch the corresponding video clip. I had great fun making the video clips and hope that you will find pleasure in them too. So now, what are we waiting for? Let's get on with our exquisite bit of flapdoodle.

Chapter 2

What's the Flap?

Now that we have had a quick introduction to piano-hinged dissections, it's time to slow down and think more carefully about them. Before we invest too much energy in finding them, we should make sure that they have a sound mathematical foundation. Also, it would be nice to develop some appreciation for the special properties that piano-hinged dissections can exhibit. Otherwise, why make such a big flap? Let's start by refining our model of piano-hinged dissections, and then explore the properties of a specific example. That will lead us to survey a variety of cyclic hingings and conclude the chapter by considering some additional properties.

The model for piano hinges. Although the examples of piano-hinged dissections in Chapter 1 may seem straightforward, we have already been careful when we defined our piano-hinging of "two-dimensional" figures. If we had assumed that each level is truly two-dimensional, i.e., with no thickness, then two pieces connected by a piano hinge would share points in common when we fold them against each other in three-dimensional space. We avoided this problem by assuming that each piece is the union of one or more prisms of thickness ϵ, for some small $\epsilon > 0$. The axis of rotation of a piano hinge will coincide with an edge on the shared boundary of two pieces in each of the two figures that the pieces form. We see a schematic of two such prisms hinged together in Figure 2.1. It is easy to rotate piece A by 180° to bring it to lie on top of piece B, as in Figure 2.2.

However, when we assume a positive thickness for the pieces, we see a further problem unfold with certain dissections. One piece may obstruct another by a tiny amount (a function of ϵ) when we rotate about one of the hinges. For example, the 2-piece hinged assemblage in Figure 2.3 will not rotate through a complete 180°, even though pieces A and B do fit quite

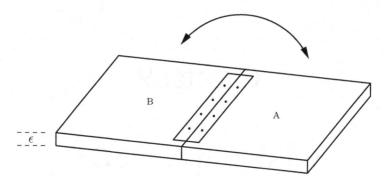

Figure 2.1. Folding two pieces from being next to each other . . .

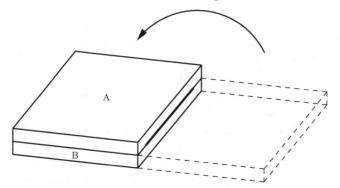

Figure 2.2. . . . to being one on top of the other.

well together. The problem is that during the rotation, the leading edge
on what is originally the top level of piece A will bump into the top of the
edge of piece B, as we see in Figure 2.4. This problem arises only because
the edges have positive thickness and would not arise if we had "pieces"
with no thickness.

We can modify such dissections by "rounding" certain crucial surfaces.
Specifically, we make the surfaces concave by carving away a small (as a
function of ϵ) amount of material from the side of a prism. For example, in
Figure 2.5, we can overcome the problem by rounding that crucial edge of
piece B to conform with the circle arc indicated by the dashed edges. The
volume of material lost due to rounding becomes a negligible fraction of
the total volume as we decrease ϵ. We call these dissections *rounded piano-
hinged*. Actually, we have already seen such a dissection in Figures 1.5 and
1.6, in which we should round certain surfaces of pieces E and F.

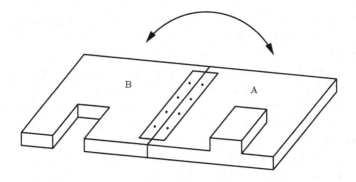

Figure 2.3. Piano-hinged assemblage with a minor obstruction.

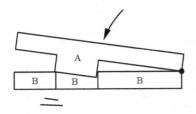

Figure 2.4. Top edge of B obstructs.

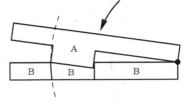

Figure 2.5. Rounded edge okay.

Now let's recap our understanding of piano-hinged dissections. We represent two-dimensional figures by shallow prisms. Each prism is separated into two levels, each of thickness ϵ, and each level is dissected into smaller prisms, each of thickness ϵ. Each piece in a piano-hinged dissection is either a smaller prism or comprises the union of such smaller prisms. In the latter case, the prisms that are unioned must be contiguous. For example, in Figure 1.5, piece E is the union of two prisms E_1 and E_2, and piece I is the union of two prisms I_1 and I_2. For a piano-hinged dissection of *one* figure to *one* other figure, the pieces must be connected together into a single assemblage by piano hinges.

It is natural to extend this definition to define a piano-hinged dissection of n figures to one figure by requiring that it consist of precisely n such assemblages. We can even dissect n figures to m figures, in which case we must use the minimum possible number of assemblages. See Figures 8.9, 9.7, 9.34, 10.6, 10.10, and 19.13 for examples of n figures to m figures, with both n and m greater than 1.

A folding model. In his *Menon* and also in his *Timaeus*, Plato described a rather simple unhinged dissection of two equal squares to a larger square. We can adapt the dissection into a nettable dissection by using the dissection on each level and hinging the eight pieces appropriately, as in Figure 2.6. Following the convention from the first chapter, I use primed letters to indicate pieces from the second copy. We see the two assemblages in a perspective view in Figure 2.7. The assemblage on the left shows how the pieces change from a small square to the left half of the large square. The assemblage on the right has a corresponding labeling of pieces but rotated around in forming the right half of the large square. This dissection has a number of nice properties, which we shall identify shortly.

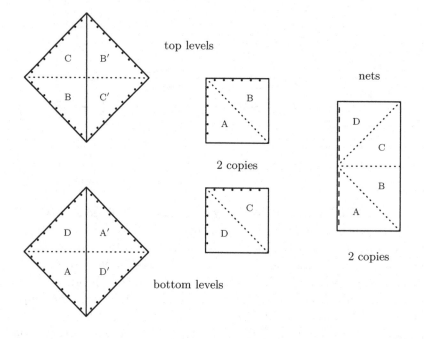

Figure 2.6. Nettable dissection of two equal squares to one. (W)

In order to lay out each assemblage of pieces in a net, I use dashed edges to indicate the hinge that connects pieces A and D. Observe that in each of the two assemblages there is one more hinge than we really need. If we remove one piano hinge from each assemblage, the pieces in the assemblage remain connected. This is a feature of cyclic hingings that we will discuss in the next section.

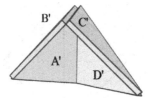

Figure 2.7. Perspective view of the assemblages for two equal squares to one.

If we make a model of this dissection out of pieces thicker than paper, we need to take care when attaching the piano hinges. Should we center the central axis of the hinge's pin between pieces A and B on the top or the bottom surface of the panels as they lay in the small square? The answer is the bottom surface. Similarly, the connection between pieces C and D in the small square should be on the top surface of C and D. In the interest of limiting notation within the figures, I do not indicate in general which surface to use. It is not hard to infer the side on which to attach the hinge, if we look at both the large square and the small squares. If you are constructing a model with real (hardware) hinges, try it first with tape to verify where the hinges go before you cut any material.

We may choose to construct models of our piano-hinged dissections out of thin panels of wood. Suppose that we cut the pieces so that on each level the grain runs uniformly in some direction when we form one of the two figures. If when we then form the other figure, the grain runs uniformly in some direction on each level, we say that the dissection is *grain-preserving*. For example, the dissection of a triangle to two rectangles (Figure 1.8) is grain-preserving.

Relatively few piano-hinged dissections seem to have this property, which comes in two variations. We say that a piano-hinged dissection is *any-grain-preserving* if we can run the grain uniformly in any direction on one level and uniformly in some other (corresponding) direction on the other level. We say that the dissection is *bi-level-grain-preserving* if there are two (orthogonal) directions such that the grain runs the same direction uniformly on both levels simultaneously.

The dissection in Figure 2.6 is not bi-level-grain-preserving, but it is any-grain-preserving, as we can see in Figure 2.8. The dissection in Figure 1.8 is both any-grain-preserving and bi-level-grain-preserving. It is bi-level-grain-preserving because we can run the grain vertically or run the

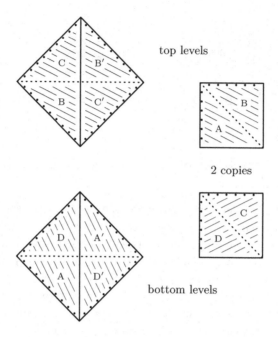

top levels

2 copies

bottom levels

Figure 2.8. Piano-hinged two equal squares to one is any-grain-preserving.

grain horizontally. It would be a bit much to identify all of the piano-hinged dissections in this book that have these grain-preserving properties; you will have to be on the lookout for them yourself. They pop up in the chapters dealing with squares and rectangles, such as Chapters 5, 6, 7, 8, 10, 12, 13, and 15. You can also identify a handful of them in the video that demonstrates the wood models.

Cyclic piano-hingings. A lovely property of some piano-hinged dissections is that they have a cyclic hinging. A dissection is *cyclicly hinged* if we can remove one of the hinges without disconnecting any of the pieces from an assemblage. This over-abundance of hinges forces various pieces to move in a neatly coordinated fashion. There are several different ways that we can cyclicly hinge pieces. For ease in characterization, we shall imagine that we have shrunk ϵ to zero. The first type of hinging, a *vertex-cyclic hinging*, has four or more pieces that touch at a common vertex, and each piece is hinged with its predecessor and successor on the cycle.

Let 2σ be the sum of the angles that meet at the vertex. If $\sigma < 180°$, then the vertex-cyclic hinging is a *cap-cyclic hinging*. We have already seen an example of a cap-cyclic hinging in Figure 2.6, in which each small square is cap-cyclicly hinged. A number of the video demonstrations feature cap-cycles. In addition to the wooden model for Figure 2.6, be sure to view the intriguing wooden model for Figure 12.6, which shows what happens when several cap-cycles interact together.

All of the piano-hinged dissections in this book that are cap-cyclicly hinged have four pieces meet at a cap-cyclic vertex. (You may wish to think about why this turns out be the case.) When we fold the cap-cycle flat in either configuration, the angle that we form will be σ. One pair of nonadjacent pieces will have an angle of α, where $\alpha < \sigma$, at the vertex, and the other pair of nonadjacent pieces will have an angle of $\sigma - \alpha$. (Again, you may wish to think about why.)

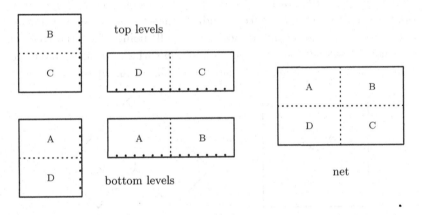

Figure 2.9. Folding dissection of ($a \times 2b$)-rectangle to ($2a \times b$)-rectangle.

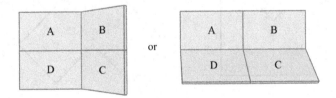

Figure 2.10. Two ways to fold a flat-cyclicly hinged set of pieces.

If $\sigma = 180°$, then the vertex-cyclic hinging is a *flat-cyclic hinging*. In Figure 2.9 we see an example of a flat-cyclic hinging, namely the 4-piece folding dissection of an $(a \times 2b)$-rectangle to a $(2a \times b)$-rectangle. Examining the net in that figure, pieces A, B, C, and D are cyclicly hinged and all meet at a point in the center. The angles at that common vertex sum to $2\sigma = 360°$. Folding along the hinges that separate piece A from B and C from D gives the $(a \times 2b)$-rectangle. We see the beginning of such a fold on the left in Figure 2.10. Folding along the hinges that separate A from D and B from C gives the $(2a \times b)$-rectangle. The beginning of this fold is on the right in the figure. Some other examples of flat-cyclic hingings are in Figures 1.11, 5.10, 6.2, 6.6, and 6.9.

If $\sigma > 180°$, then the vertex-cyclic hinging is a *saddle-cyclic hinging*. This cycle will be at a concave vertex in the boundary of each target figure of the dissection. Its name derives from the shape of the surface as it moves from one configuration to the other. The assemblage will open from a concave angle to a surface with a saddle point and then close to a concave angle on the other side of the surface. At a saddle point (of a surface shaped like a horse's saddle), one cross-section of the surface attains a maximum while another cross-section of the surface attains a minimum.

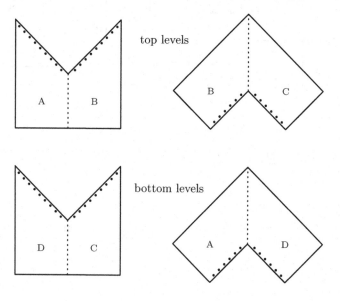

Figure 2.11.   Saddle-cyclicly hinged mitre to gnomon. (C)

As an example, we see the piano-hinged dissection of a mitre to a gnomon in Figure 2.11. The *mitre* is what remains from a square when we cut away an isosceles right triangle whose hypotenuse coincides with a

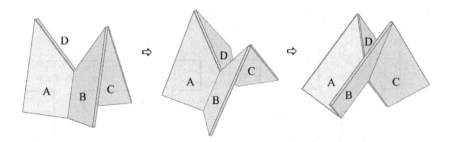

Figure 2.12. Saddle-cyclicly folding a mitre to a gnomon.

side of the square. Sam Loyd (*Inquirer*, 1901a) featured a mitre in what turned out to be an erroneous dissection. The *gnomon* is what remains from a square when we cut away a square of one fourth the area from a corner of the larger square. Philip Kelland (1855) featured the gnomon in a whole set of dissections. Both the mitre and the gnomon have an angle of $\sigma = 270°$. On each level we cut the figure in two, splitting these large angles and thus creating four pieces that meet at angles whose total is 540°. We see how to fold from the mitre to the gnomon in Figure 2.12. The video demonstrations of card stock models include not only the saddle-cyclic model for Figure 2.11 but also saddle-cyclic models for Figures 17.9 and 17.11.

The remaining types of cyclic hingings are not vertex-cyclic. The first of these is *tube-cyclic*, in which the cycle involves a sequence of pieces that do not share a common vertex. An example is the dissection of the T-pentomino to the U-pentomino in Figure 2.13. The U-pentomino appears upside down in the figure, because that makes it easier to see how the folding works. Pentominoes are figures formed by gluing five congruent squares edge to edge. We shall study them in greater detail in Chapters 11 and 12. In Figure 2.14 we see that pieces B, C, D, and E form a cycle. As we unfold and then refold the assemblage, we take the center column of the T and distribute it to the two arms of the U.

To see tube-cyclic hingings in action, view the video demonstrations of the card stock model for Figure 5.1 and the wooden model for Figure 7.7. Tube-cyclic hingings are so nifty—let's take time out to find another one, this time between a T-pentomino and a V-pentomino. The V-pentomino has a column like the T-pentomino, but instead of arms of equal length on both sides, it has no arm on the left and an arm of length 2 on the right. (It looks like a V if we rotate it 135° counterclockwise.)

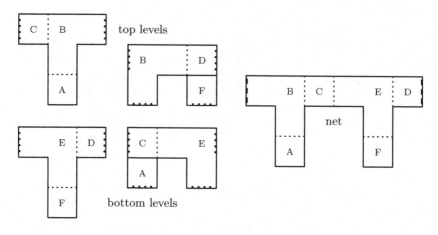

Figure 2.13. Tube-cyclicly hinged T-pentomino to U-pentomino. **(C)**

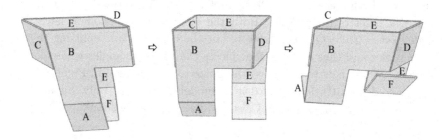

Figure 2.14. Tube-cyclicly folding a T-pentomino to a U-pentomino.

Puzzle 2.1. *Find a 4-piece tube-cyclicly hinged dissection of a T-pentomino to a V-pentomino.*

Another type of tube-cyclic hinging results when we take a set of cap-cyclic hinged faces and cut off portions of the pieces next to the shared vertex. We will see an example of this in Figure 15.3, in which faces A, B, C, and D would be cap-cyclicly hinged if they extended sufficiently far to the left in the large cross.

The remaining type of non-vertex-cyclic hinging is *leaf-cyclic*. In a leaf-cyclic hinging, four pieces share a common edge and connect to it like the pages of a book connect to its binding. The pieces flap around this spine like a butterfly's wings flap around its thin body. I shall designate the hinges for the leaf-cyclic hinging by two parallel rows of dots that are offset from each other and separated by a thin line. To see leaf-cyclic hingings in action, view the video demonstrations of card stock models for Figures 5.4 and 5.8 and the wooden model for Figure 16.18.

Other properties. Several more properties are worth mentioning. The first is the lovely "inside-out" property. Suppose that each piece is on only one level, so that the piece has two primary surfaces. When we assemble the pieces to form one figure, one surface is on the outside, and the other surface is on the inside (i.e., hidden from view). If, after we fold the pieces to form the other figure, all of the surfaces that were on the outside are now on the inside, and vice versa, we say that the dissection has the *inside-out* property. The saddle-cyclicly hinged dissection of a mitre to a gnomon, which have just seen in Figures 2.11 and 2.12, has this property, as you can see in the video demonstration of the card stock model of this dissection. The inside-out property is reminiscent of the situation in origami in which the paper that we fold has a different color on each side.

The reverse of the inside-out property is the *exterior-preserving* property, in which those surfaces exposed to the exterior stay on the exterior, and those on the interior stay on the interior. Examples of dissections with this property are in Figures 1.3, 1.5, 1.11, 2.6, and 2.13.

Cap-cyclic and tube-cyclic hingings are mutually exclusive with the inside-out property. Similarly, saddle-cyclic and leaf-cyclic hingings are mutually exclusive with the exterior-preserving property. Interestingly, there are two ways to accomplish a flat-cyclic hinging. One is consistent with the inside-out property, and the other is consistent with the exterior-preserving property. If we interchange the top and bottom levels of the rectangle in the dissection in Figure 1.11, then the resulting hinged dissection has the inside-out property rather than the exterior-preserving property.

With all of these properties, what more could we want? Lots and lots of piano-hinged dissections, that's what! To obtain them, we will first backtrack to other types of hinged dissections. Then we'll see how to produce piano-hinged dissections in mass quantities—all coming up in the next two chapters.

Manuscript 1

The "Lost" Manuscript of Ernest Irving Freese

With exuberance tempered by the hint of reality, Los Angeles architect Ernest Irving Freese announced the completion of his opus on geometric dissections in a letter to his friend Dorman Luke on July 20, 1957:

> For the past 4 months I've let everything go hang — work, friends, correspondence, or what have you — BUT, it's finished — the first book on Geometric Transformations (Dissective Geometry) ever produced. 200 plates comprising about 400 original examples. Cost $28.$\underline{^{00}}$ just to get a blueprint copy of the dwgs. Well, it's off my chest — but who will publish it? Probably <u>nobody</u>!

Three months to the day later, Freese died of coronary arteriosclerosis and was soon cremated. Aside from copies of a dozen plates sent to friends, his book vanished. It was entombed for the next forty-five years in the house that he had converted from a Prohibition-era shack to a studio, then enlarged as his first family moved in during the depths of the Great Depression, and in which he lived out the remainder of his life with a second family.

Those few who knew of the manuscript tried to contact the widow, Winifred Freese, to encourage her to release it for publication. She never responded, so that by 1965 people had given up trying to contact her. Thirty years later, when I obtained copies of correspondence of the by-then deceased Harry Lindgren and identified the address 6247 Pine Crest Drive, I was able to pick up the search. Although Mrs. Freese had also passed away, her son Bill still lived in that same house and answered my letter.

Unfortunately, Bill Freese did not know if any of his father's work on dissections had survived. When Bill was hospitalized with cancer in 2002, his cousin and designated heir Vanessa Kibbe started to clean out the house. Finding an unopened letter from me to Bill alerted her to the existence and importance of the manuscript, which she then located several months later.

GEOMETRIC TRANSFORMATIONS *
BY
ERNEST IRVING FREESE
....

*A GRAPHIC RECORD of explorations and discoveries
in the diversional domain of DISSECTIVE GEOMETRY*
....

Figure M1.1.   From the title page of Ernest Freese's manuscript.

What a thrill, to bring this legendary manuscript back from the brink of oblivion!

And how fascinating to examine this time capsule from the 1950s! The inventiveness that Freese brought to his geometric explorations is impressive. There are some remarkable gems, such as the 6-piece dissection of a regular dodecagon to a regular hexagon, or the 9-piece dissection of an unusual cross to a square, which is a double application of the completing-the-tessellation technique. The range of new dissection puzzles is breathtaking, as we shall see subsequently in parts 2, 4, and 5 of this brief survey.

The graphic beauty of the pages is stunning: elegantly laid out drawings, with meticulously dotted dimensions and selective thickening of well-chosen line segments and letters, and the lovely text, with stylized, slanted lettering and numbers. What a feast for the eyes! Freese laid out each plate in pencil on an $8\frac{1}{2}$-by-11-inch sheet of paper, often with narrow margins.

A captivating example is Plate 86 (Figure M1.2), in which Freese artfully arranged three new dissections. The nifty dissection of three hexagrams to one at the top of the page preceded the discovery by Harry Lindgren (1964a). Freese's dissection of four hexagrams to one in the middle of the page fell victim later to a less regular 11-piece dissection by Robert Reid. Finally, Freese's symmetric perfection at the bottom of the page, a dodecagram to a hexagram, was handily overtaken by Lindgren (1964b). Also visible on the lower left are the identification strings of Freese's cryptic cataloging scheme.

The manuscript has its problems. First, there is a frustrating lack of citations. Freese was aware of at least some but probably not all of the dissection work that preceded his. For example, did he know about the techniques of Harry Hart (1877) for transforming two similar copies of one figure to another? Or did he rediscover his limited version of them? Second, there is no introduction to the manuscript. We can only infer how Freese hoped readers would receive his work and what he thought of it. Third, Freese named various techniques but neither defined nor explained them. Finally, he made claims that are either wrong or for which neither he nor anyone else before or since has provided any substantiation.

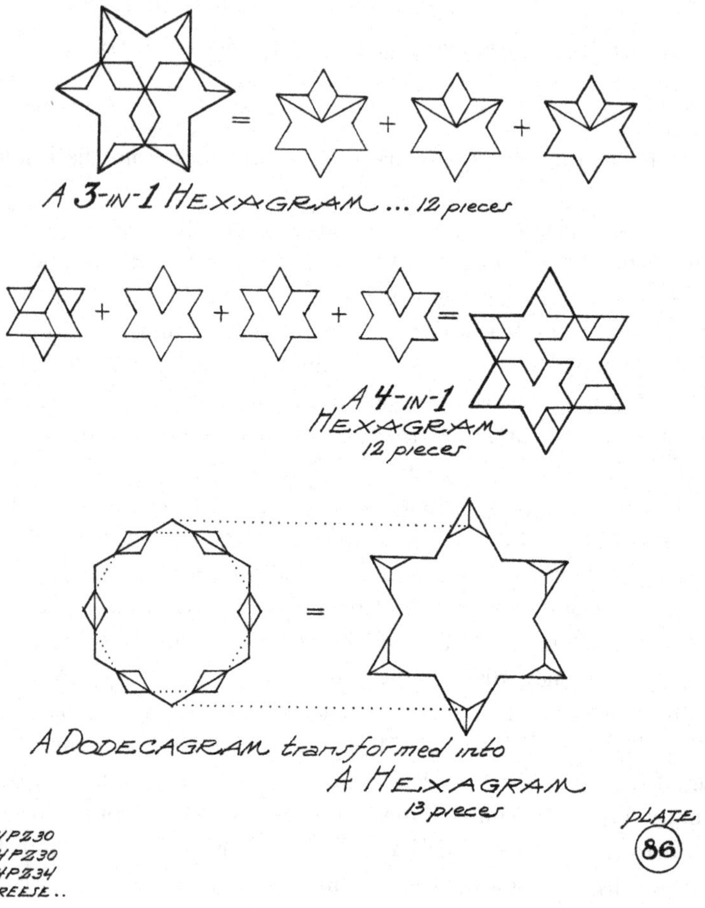

Figure M1.2.   Ernest Freese's Plate 86.

No matter what the shortcomings of the work, it is nonetheless a unique document that has the power to entrance and inspire those who are drawn into it, as we see with the lower half of Plate 95 (Figure M1.3).

8 OCTAGONS MAKE 1 OCTAGON

B2.PR7D.S1
D2.PR7F.S1
..FREESE..

24 pieces

PLATE
95

Figure M1.3.   Lower half of Ernest Freese's Plate 95.

Who was Ernest Irving Freese, and how had he come to craft his mesmerizing manuscript? Born on February 5, 1886 in Minneapolis, Minnesota, Ernest was the older of two children of William Henry Freese, a cooper who had come from Maine, and Bertha Reeves, who died when Ernest was young. He quit school after the eighth grade and entered the employ of an architect, working as a draftsman and as a construction superintendent.

From 1911 to 1923, Freese was associated with the Los Angeles firm of Hunt and Burns and claimed to be in complete charge of all of their architectural and structural design. He claimed to have personally designed the Automobile Club and the Southwest Museum, both in Los Angeles, and the Scottish Rites Temple in Santa Fe, New Mexico. Subsequently, he practiced architecture independently, aside from being the key man on army and navy projects in the First and Second World Wars.

In a 1948 response to a query from the A. N. Marquis Company, Freese gave the following self-characterization:

> Belong to no society, association, club, church or political organization: in fact am hated most cordially by most of them! Strictly a "lone wolf" who never knows which side of the fence his bread is buttered on, nor cares.

After his formal schooling, Freese continued his education through self-study and writing. He studied geometry, trigonometry, and calculus, producing exquisitely drafted notebooks. Over three decades, starting in 1912,

Freese published more than a hundred articles on topics like architectural drafting, house design, structural design, Hawaiian culture, and motorcycling. Some of the articles formed series, including a five-part series on perspective projection (1929b), a 22-part series on architectural drafting (1929a), and a 14-part series on structural design (1939). Taken as a whole, his articles demonstrated singular initiative and talent in proceeding with writing projects. With this substantial record under his belt, it is less surprising that he would create his remarkable manuscript late in his life.

When Freese embarked on a project, he did so with energy, verve, and often humor. In Figure M1.4, we see the tongue-in-cheek "biography" that he supplied for the March 1930 issue of *Pencil Points*. So suitably warned, grab your compass, your triangles, and your T-square, and try to keep up as we make a quick dash through this man's intriguing manuscript!

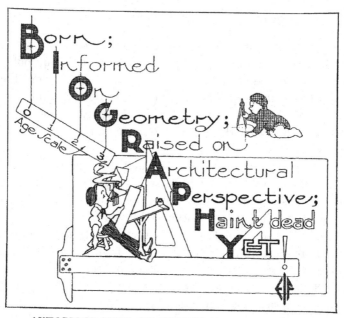

AUTOBIOGRAPHY OF ERNEST IRVING FREESE

Figure M1.4.   Ernest Freese's tongue-in-cheek "BIOGRAPHY".

Chapter 3

Swingers and Twisters

Now that we have an idea of what piano-hinged dissections are like, let's
see what they are not like. In this chapter we will contrast them with
hinged dissections that swing or twist. A lovely example is the dissection
of an equilateral triangle to a square that is at the top of Figure 3.1. The
dissection is *swing-hinged* because we can link its pieces together with *swing
hinges*, so that when we swing them one way on these hinges, the pieces

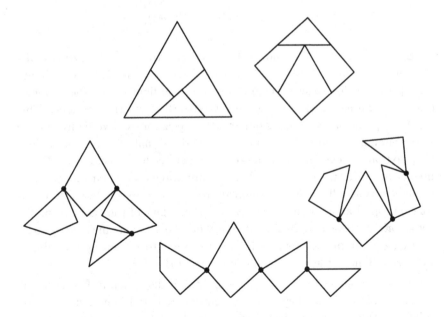

Figure 3.1. Swing-hinged dissection of a triangle to a square. (W)

form one figure, and when we swing them around in another way, the
pieces form the other figure. Henry Dudeney (1907) was the first to point
out this eye-catching property, five years after he had originally published
(in *Dispatch*, 1902b) the 4-piece triangle-to-square dissection, perhaps after
learning of it from Charles McElroy (Dudeney, *Dispatch*, 1902a). Note that
swing-hinged dissections have pieces on just one level, because swing hinges
do not move pieces from one level to another.

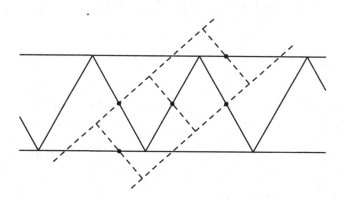

Figure 3.2.   Crossposition of triangles and squares.

Harry Lindgren (1951) showed how to derive the triangle-to-square dissection by *crossposing* two *twin-strips*, or *T-strips*, as in Figure 3.2. Create
a strip by packing together copies of one of the figures (say the triangle)
one after another, with each figure rotated 180° from the previous. This
creates a point of rotational symmetry between consecutive figures in the
strip. Similarly create a strip of squares, and lay that strip across the first
strip so that the overlapped area is either equal to the area of one of the figures or equal to twice its area. Furthermore, force each point of rotational
symmetry that falls in the overlapped area, as indicated by dots, to be either on top of another such point or on a boundary of the other strip. We
get a *plain strip*, or *P-strip*, if we do not require each figure in the strip to be
rotated 180° from the previous figure. However, to get swing-hingeability
we normally need the additional symmetry of the T-strip.

While people have viewed the swing-hinged dissection in Figure 3.1 as
a remarkable curiosity for almost a century, it is only one out of a vast
collection that I explored in my second book (2002). With such a wealth
of examples, we must acknowledge a question that begs for resolution:

Given any pair of figures that are of equal area and bounded by straight line segments, is it always possible to find a swing-hinged dissection of them?

One of the most elementary of the swing-hinged dissections is the 2-piece dissection of a mitre to a gnomon in Figure 3.3. Too simple to be amazing, it will nonetheless play a pivotal role soon when we use it to illustrate a technique that converts swing-hinged dissections to twist-hinged dissections.

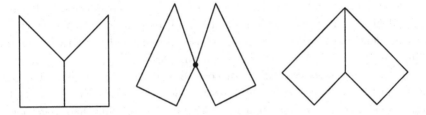

Figure 3.3.   Swing-hinged dissection of a mitre to a gnomon.

A *twist hinge* is a hinge that uses a point of rotation on the interior of an edge shared by two pieces. With a twist hinge, we flip over one piece relative to the other, using rotation by 180° through the third dimension. Pieces A and B (with exaggerated thickness) are twist-hinged together in Figure 3.4.

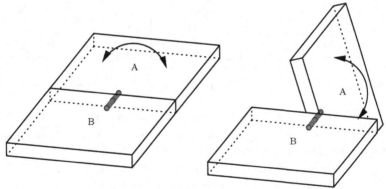

Figure 3.4.   A twist hinge for pieces A and B.

At first this seemed to be excessively restrictive, until I discovered the many examples that I included in my book (2002). Prior to that work, only three simple examples, by Rubik (1983), by Esser (1985), and by Lurker (1984), had been known. Yet again, my second book (2002) identified a second outstanding open problem:

> Given any pair of figures that are of equal area and bounded by straight line segments, is it always possible to find a twist-hinged dissection of them?

It is remarkable that, using appropriate surgery on the pieces, we can often convert a swing-hinged dissection into one that is twist-hinged. For example, let's consider the dissection in Figure 3.3. We produce a twist-hinged dissection from it as follows: For each piece adjacent to the swing hinge, identify an isosceles triangle whose apex coincides with the swing hinge and each of whose two equal-length sides coincides with a portion of the piece's boundary that is adjacent to the swing hinge. The dashed edges in Figure 3.5 indicate the bases for the two isosceles triangles. Cut the isosceles triangles from their respective pieces and glue them together, giving a third piece. Finally, position twist hinges at the midpoints of the bases of the isosceles triangles. The result is the 3-piece twist-hinged dissection in Figure 3.6. Note the special marks on pieces that I turn over an odd number of times as I twist the assemblage from the mitre to the gnomon: an "*" in the mitre and a "⋆" in the gnomon.

  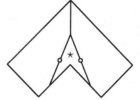

Figure 3.5. Isosceles.   Figure 3.6. Twist-hinged mitre to a gnomon.

The technique for converting swing-hinged dissections requires that the dissection have binary hinges (connecting two pieces) and that the hinged pieces are "hinge-snug." Two pieces connected by a hinge are *hinge-snug* if they are adjacent along different line segments in each of the target figures, and each such line segment has one endpoint at the hinge. We can then replace each swing hinge with a new piece and two twist hinges. The new piece will be the union of two isosceles triangles, one carved from each piece attached to the swing hinge.

Using this technique, I discovered the 7-piece twist-hinged dissection of a triangle to a square by converting the 4-piece swing-hinged dissection

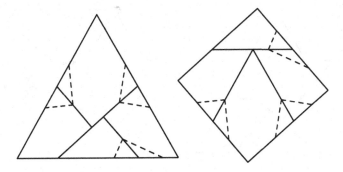

Figure 3.7. Adding isosceles triangles to the triangle-to-square dissection.

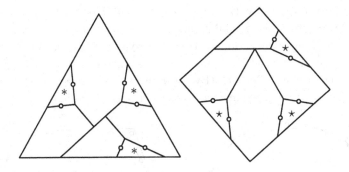

Figure 3.8. Twist-hinged dissection of a triangle to a square. (W)

in Figure 3.1. Figure 3.7 shows the cuts in the triangle and the square, plus dashed edges to indicate the bases of the isosceles triangles adjacent to each hinge point. In Figure 3.8, I have merged pairs of isosceles triangles together, giving a 7-piece twist-hinged dissection. The intermediate configurations in Figure 3.9 suggest a sequence of movements that convert the triangle to the square. On the left, we flip up the lower left corner of the triangle, using a pair of twists. Then on the right, we similarly flip up the lower right corner of the triangle. The conversion simulates each swing hinge by two twist hinges and an extra piece, as described fully in my second book. Don't miss the demonstration of the model for this dissection in the video for the wooden models.

We can find twist-hinged dissections without deriving them from swing-hinged dissections. For example, there is a truly extraordinary family of dissections. For any $p > 2$, there is a $(2p + 1)$-piece twist-hinged dissection

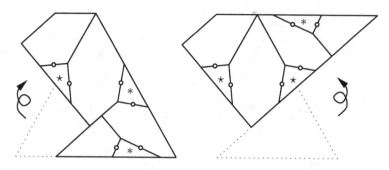

Figure 3.9. Intermediate configurations for a twist-hinged triangle to a square.

of a $\{2p\}$ to a $\{p\}$. (By $\{p\}$ we denote a regular polygon with p sides.) These elegant dissections exhibit rotational symmetry about their centers. Such a twist-hinged dissection, of a decagon to a pentagon, is in Figure 3.10. Furthermore, there is a $(2p+1)$-piece twist-hinged dissection of a $\{p/q\}$ to a $\{p\}$, whenever $p \geq 3q - 1$. (By $\{p/q\}$ we denote a regular star with p points and every point connected to the qth point from it.)

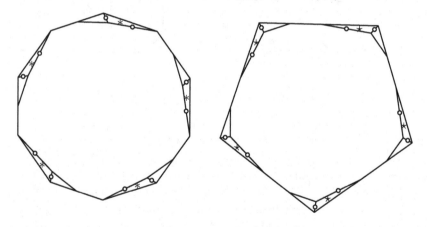

Figure 3.10. Twist-hinged decagon to a pentagon.

Yet these are not the only infinite classes of twist-hinged dissections, as the next example illustrates. It's a bit involved, so you who were previously unfamiliar with twist-hinged dissections may wish to skip the rest of this chapter on a first reading. In July 2001, Wolfgang Stöcher of Austria made a striking discovery that enabled me to find another class. For any $n \geq 3$, Stöcher found a symmetric method to dissect a $\{3n\}$ to an $\{n\}$, using $4n + 1$ pieces. Within a couple of days, he and I had independently discovered how to reduce the number of pieces to $3n + 1$. He went on to reduce the number of pieces further, to $\lceil 5n/2 \rceil + 1$ pieces, while I went on to discover a $(4n + 1)$-piece swing- and twist-hinged variation that is hinge-snug. The latter variation led me to a $(5n + 1)$-piece twist-hinged dissection.

Of course, when $n = 4$, this is not particularly impressive, because I had already found a 9-piece twist-hinged dissection of a dodecagon to a square. However, the method that I have adapted from Stöcher's approach works for any $n > 2$. Even for $n = 3$, it makes an advance. Although Robert Reid and Gavin Theobald had found unhingeable 8-piece dissections and Anton Hanegraaf had found an 11-piece hingeable dissection with some swing hinges that are not hinge-snug (see my previous books), no one had a clue as to how to produce a twist-hinged dissection.

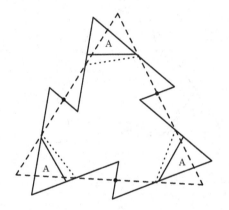

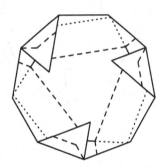

Figure 3.11. Enneagon element.   Figure 3.12. Elementary surgery.

Let's start with Stöcher's basic dissection and work our way step by step to my twist-hinged version, using the case of $n = 3$ (an enneagon to a triangle) as our example. Stöcher first cut a $\{3n\}$ to produce an element with rotational symmetry and sides that are parallel to the sides of an $\{n\}$. He cut n acute triangles, each containing a side of the $\{3n\}$ framed by angles of $240°/n$ and $120°/n$. For $n = 3$, these angles are $80°$ and $40°$. Stöcher flipped each acute triangle over and placed it flush against the next side of the $\{3n\}$, producing an irregular $\{3n\}$ with $2n$ angles of $(1 - 2/n)180°$ and n angles of $(1 + 2/n)180°$. For $n = 3$, we see the irregular enneagon in the solid edges of Figure 3.11, with each acute triangle labeled by an A.

Stöcher overlaid an $\{n\}$ of area equal to the $\{3n\}$, so that the centers coincided and each edge of the $\{n\}$ bisected a short edge of the irregular $\{3n\}$. For $n = 3$, the dashed edges in Figure 3.11 illustrate this corresponding equilateral triangle. The cuts induced by the dashed and solid edges produce the $(4n+1)$-piece dissection that Stöcher discovered. We can refine this dissection to $3n+1$ pieces by finding the points near each vertex of the $\{n\}$ where the boundary of the $\{n\}$ crosses the boundary of the irregular $\{3n\}$. If we match the nearest pairs of points (dotted lines) and cut along the corresponding line segments, we avoid an extra line crossing, giving the dissection in Figure 3.13. This has three acute triangles and also three obtuse triangles, each of the latter labeled with a B. Stöcher and I found this improvement independently.

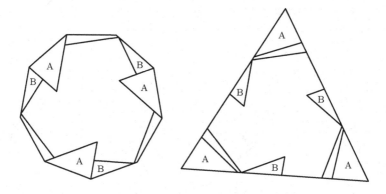

Figure 3.13. Unhingeable enneagon to a triangle.

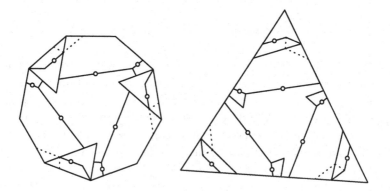

Figure 3.14. Swing- and twist-hinged enneagon to a triangle.

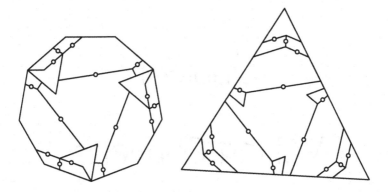

Figure 3.15. Twist-hinged enneagon to a triangle.

We can use swing or twist hinges on some of the pieces in Figure 3.13. To make the dissection completely swing- and twist-hingeable, slice n isosceles trapezoids off the large central piece, and slice an isosceles triangle off each of the n obtuse triangles. Then, glue an isosceles trapezoid and an isosceles triangle to each piece cut in refining the dissection to $3n+1$ pieces. Dashed edges indicate the cuts, and dotted edges indicate the gluings in Figure 3.12. To connect the acute triangles to the glued-up pieces via swing hinges, cut an isosceles trapezoid out of each acute triangle. The resulting $(4n + 1)$-piece dissection in Figure 3.14 is swing- and twist-hingeable. Dotted line segments identify the position of the three swing hinges in the enneagon and in the equilateral triangle. Each swing hinge is between an isosceles trapezoid and a glued-up piece and is hinge-snug.

Finally, we convert the hinge-snug swing hinges to twist hinges, yielding a $(5n + 1)$-piece twist-hinged dissection. For our example with $n = 3$, we get the 16-piece twist-hinged dissection in Figure 3.15. Sixteen pieces, or $5n + 1$ in general, may seem like a lot of pieces. If that bothers you, try to do better—it's not so easy. Most readers may find it sufficiently challenging just to follow the twists and turns in the development of this dissection!

Now that you have picked up the beat, are you ready to stay out all night with the swingers and twisters? Don't say no, or you'll be missing a whole lot more. Just as we can convert swing-hinged dissections to twist-hinged ones, we will see in the next chapter how to convert twist-hinged dissections to piano-hinged ones. So after swingers and twisters, brace yourself—for flappers!

Chapter 4

Willing Converts

As we have seen in the last chapter, it's a crazy world out there, with all manner of strange goings-on: dissections swinging with wild abandon, swingers transforming into twisters, whole families twisted beyond recognition, and puzzles that that defy our best efforts at solution. We could even have seen a "hex-change operation" (such as an exotic twisted dissection of a triangle to a hexagon), but the author thought it better to hush up those things.

Is there some way that we can impose an order on this chaos? That we can call these swingers and twisters to account and bring them back into the fold? How can we turn this situation around, cutting through established convention to tame a world that threatens at every opportunity to become unhinged? In response, I will show how to sweep through the twisted landscape and force conversions—by scissors (and tape)—to lay a foundation for the new era of piano-hinged dissections.

Remarkably, we can simulate a twist hinge by three piano hinges, introducing two new pieces in the process. Let's see how this idea works when we convert the 2-piece twist-hinged dissection of an ellipse to a heart, which is the two-dimensional analog of William Esser's (1985) dissection of an ellipsoid to a heart-shaped object.

To get the twist-hinged dissection in Figure 4.1, we slice the ellipse through its center along a line diagonal to its axes. Then, we can convert between the ellipse and the heart shape by rotating one piece 180° relative to the other. A small open circle indicates the position of a twist hinge, whose axis of rotation is perpendicular to the diagonal cut.

The 4-piece piano-hinged dissection is in Figure 4.2. The lower level of piece B (in the ellipse) shares a piano hinge with the upper level of piece A.

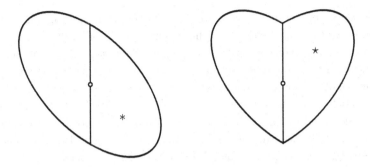

Figure 4.1. Twist-hinged dissection of an ellipse to a heart.

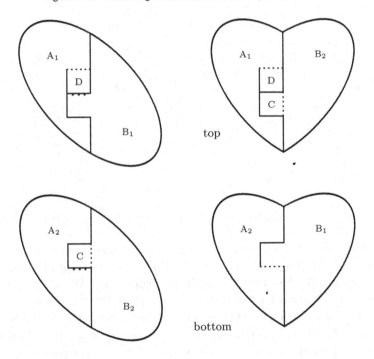

Figure 4.2. Rounded piano-hinged dissection of an ellipse to a heart. (C)

The axis of the piano hinge lines up with the axis of rotation of the twist hinge, enabling the piano hinge to simulate the action of the twist hinge. To be able to perform the rotation along the piano hinge, I have cut pieces C and D out of piece A. Once we fold them out of the way, we can rotate piece B around, and then fold pieces C and D back into position.

We see a perspective view of the assemblage in mid-fold in Figure 4.3. At the same time, you can enjoy the demonstration of the model for this dissection on my video of card stock models. This piano-hinged dissection is rounded. To be able to get out of the way, piece D must be hinged on the top surface of the top level of the ellipse, and piece C must be hinged on the bottom surface of the bottom level of the ellipse. Without the rounding, these pieces would drag against pieces B and A, respectively, as we fold D and C out of the way in the ellipse. Also, piece B would drag a trailing edge against piece A as it is folded into position in the heart.

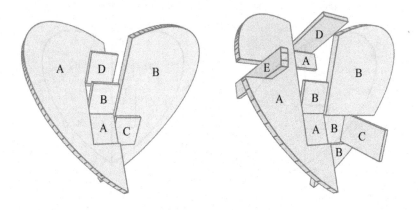

Figure 4.3. Rounded ellipse to heart.   Figure 4.4. Nonrounded model.

It is possible to avoid the rounding if we adopt a more elaborate scheme (Figure 4.5) that uses one more piece. Elongate pieces C and D, and also extend each into the other level, piano-hinging piece D on the bottom surface of the bottom level of the ellipse and piano-hinging piece C on the top surface of the top level of the ellipse. To avoid having piece B interfere with piece A, cut that portion of piece A away as a separate piece E, which we handle in much the same way as piece C. We see a perspective view of the assemblage in mid-fold in Figure 4.4.

As appealing as it is to eliminate rounding, it is not clear that we can use the more complicated scheme when converting a dissection with several twist hinges. Moreover, when we consider twist-hinged dissections of prisms of thickness 2ϵ, we need to round some surfaces for all but the simplest dissections. While Figure 4.1 requires no rounding, the twist-hinged dissections in Figures 3.6, 3.8, 3.10, and 3.15 do. We will thus revert to finding rounded piano-hinged dissections for the other twist-hinged dissec-

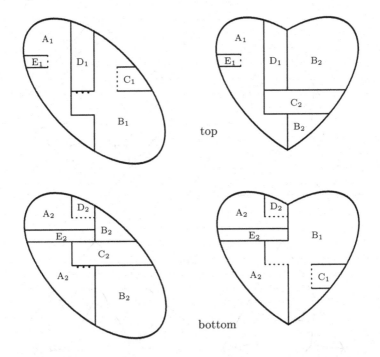

Figure 4.5. Piano-hinged dissection of an ellipse to a heart—no rounding!. (W)

tions that we wish to convert. Taken as a group, they raise hope for a positive answer to our third outstanding open problem:

> Given any pair of figures that are of equal area and bounded by straight line segments, is it always possible to find a rounded piano-hinged dissection of them?

Although we now know how to simulate each twist hinge by three piano hinges, we can often do better than to introduce two new pieces per twist hinge. We see in Figure 4.2 that piece C flips out of the way of piece A, and piece D flips out of the way of piece B, during the rotation of pieces A and B. If a piece in a twist-hinged dissection has more than one twist hinge incident on it, then we may be able to introduce just one piece that when flipped out of the way avoids all major obstructions of that piece with others. If we can do this for all pieces in the twist-hinged dissection, then our conversion will just double the number of pieces.

This is what happens when we convert the twist-hinged dissection of a triangle to a square in Figure 3.8. (Since this conversion is a bit involved,

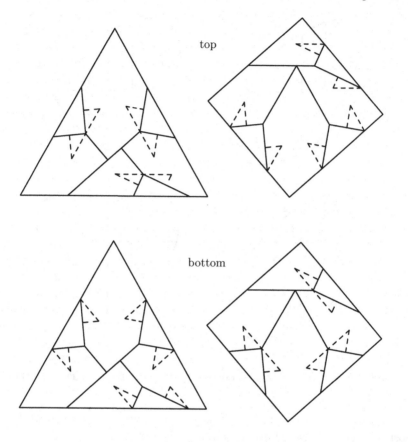

Figure 4.6. Derivation for piano-hinged dissection of a triangle to a square.

you may wish to skip the rest of the chapter on a first reading.) In Figure 4.6, we see the dissection of Figure 3.8 reproduced for both the top and the bottom levels. For each right triangle, we have smaller right triangles incident on each of the two legs, shown with dashed edges. If we remove these smaller right triangles from the large pieces, then we will be able to rotate the pieces around on axes corresponding to the legs of the small right triangles that are perpendicular to legs of the original right triangles.

In Figure 4.7, we piano-hinge piece C, corresponding to the leftmost right triangle in Figure 3.8, to pieces B and F. We piano-hinge piece D to piece C so that we can flip D out of the way. We piano-hinge piece A to B, so that we can flip A out of the way and then rotate B relative to C. On the other side of piece C, we piano-hinge piece E to piece F, so that

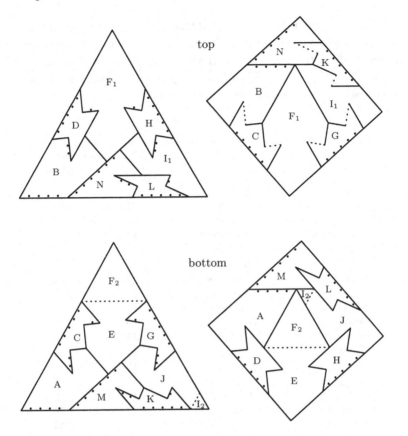

Figure 4.7. Rounded piano-hinged dissection of a triangle to a square. (W)

we can flip E out of the way and then rotate F relative to C. Note that the piano hinge for piece E will also allow us to flip E out of the way of G. Continuing this approach gives the complete hinging in Figure 4.7. The number of pieces is fourteen, twice the number used in the corresponding twist-hinged dissection.

The sequence of steps to convert the piano-hinged triangle to the square mimics the steps to convert the twist-hinged triangle to the square. First, rotate pieces B and F about piece C, as suggested above. Then, successively perform analogous operations for piece G and then for piece K.

This piano-hinged dissection is rounded. The amount of rounding is made worse by my choice of triangles, rather than rectangles, for the pieces added in Figure 4.6. When I made several models out of 3/16-inch thick wood, I compromised by using blunted triangles, as you can see in the video for wooden models. The rounding is still necessary, but not so obvious, and is in fact hidden from view when the assemblage is folded up into either the triangle or the square. With fourteen pieces, the model is a handful to demonstrate, as readers can see on the video!

First figure	Second figure	Number of pieces (twist-)	(piano-)
pentagon	triangle	12	24
pentagon	square	11	23
pentagram	square	13	27
pentagram	pentagon	11	23
hexagon	triangle	7	15
hexagon	square	6	13
hexagon	pentagon	13	27
hexagon	pentagram	15	31
hexagram	square	13	28
hexagram	hexagon	9	18
heptagon	triangle	15	32
heptagon	square	10	22
heptagon	hexagon	15	31
{7/2}	heptagon	15	31
octagon	triangle	12	24
octagon	square	9	19
octagon	pentagon	16	34
{8/2}	square	11	22
{8/3}	square	12	25
{8/3}	hexagon	14	31
enneagon	triangle	16	36
decagon	triangle	14	29
decagon	square	13	27
decagon	pentagon	11	23
dodecagon	hexagon	9	18
{12/2}	square	15	33
{12/2}	hexagon	13	26
Greek Cross	triangle	8	16
Greek Cross	dodecagon	9	19
Latin Cross	triangle	8	16
Latin Cross	square	11	22

Table 4.1. Selected dissections with conversions to rounded piano-hinged.

Aside from being a handful, the model works just fine. However, one problem still remains: Some of the piano hinges are relatively short, concentrating the torque when folding from one figure to the other. Thus, a piano-hinged dissection that does not derive from the conversion of a twist-hinged dissection may well be preferable to one that does, assuming that it requires no additional pieces.

For readers who do not find their favorite combinations of figures in this book, Table 4.1 lists examples of piano-hinged dissections that we can derive via conversions. Some of the twist-hinged dissections appear in my second book (2002) and on its web update pages. See my webpage http://www.cs.purdue.edu/homes/gnf/book3/table4.1.html for a more complete discussion and derivation of these results.

In converting certain twist-hinged dissections, we must split some pieces into more than two pieces when they are adjacent to more than two twist hinges. This explains why some entries in the rightmost column of Table 4.1 are more than double the corresponding entry in the preceding column. Again, see the above-mentioned webpage for further explanation. Even though these piano-hinged dissections do involve a lot of pieces, are we not ready, and willing, to pay that price to bring them into the fold?

In the remainder of this book, I will give piano-hinged dissections that do not come exclusively from conversions. Applying the conversion technique is relatively straightforward, and we do not need to see repeated applications of it. This approach mirrors the approach in my previous book of not identifying a twist-hinged dissection if it comes from merely converting a hinge-snug swing-hinged dissection. The dissections of some figures that are easy in swing-hinged or twist-hinged forms become more challenging for piano-hinged. Be ready to handle combinations of figures that you have not seen before—just maybe they'll make *you* flip!

Manuscript 2

Freese's Unique Point of View

Ernest Freese had a most original approach to the field of geometric dissections. He created some simply delightful dissection puzzles, ones that no one else has thought to work on, as we shall see in the next few pages.

Freese was the first person to propose dissecting several different polygons of equal area to yet another figure. Probably the simplest such example is of an equilateral triangle and a square to a hexagon, which we see in Plate 74 (Figure M2.1). How clever to dissect the square into a triangle, which he then combined with the other triangle to make the hexagon!

A second general type of puzzle that Freese proposed is the dissection of different-sized copies of the same geometric figure to a larger version of that figure. He preferred two different variations, one in which the side lengths of the figures are in arithmetic progression, and the second of which has areas in arithmetic progression. Plate 21 (Figure M2.2) showcases two dissections of the first type. The first dissection features equilateral triangles whose side lengths are in the ratio 1 : 3 : 5 : 7. The second dissection features triangles whose side lengths are in the ratio 1 : 2 : 3 : 4 : 5 : 6.

Both dissections rely on the existence of an appropriate length in a triangular grid. Sketch the resulting large triangle for each dissection on such a grid, to verify that the dissections are indeed correct. In the first dissection, the side length of the large triangle is $\sqrt{84}$, which is the length of the third side of a (nonequilateral) triangle that has an angle of 120° sandwiched between sides of lengths 2 and 8. In the second dissection, the desired side length is $\sqrt{91}$, which is the third side of a triangle that has an angle of 120° between sides of lengths 5 and 6. As it turns out, I have been able to improve on each of these dissections by one piece. I include the improvement for the first dissection in Figure M5.2 (page 240).

Freese used the technique of completing the tessellation to design a number of dissections. In that method, he added the same additional figure to each of two figures that he wanted to dissect and then formed two tessellations, each based on one of the pairs of figures. He used this technique for each of the two dissections on Plate 160 (Figure M2.3). However, for the second dissection, he used a double application of the technique: He added an octagon to the fancy cross and added a square of area equivalent to that of the octagon to the square of area equivalent to that of the cross. What a cunning dissection and conclusion to this section!

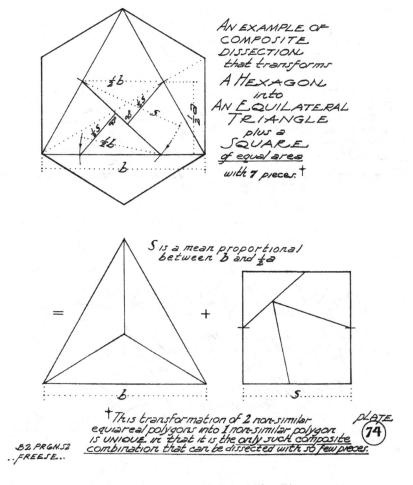

Figure M2.1.   Ernest Freese's Plate 74.

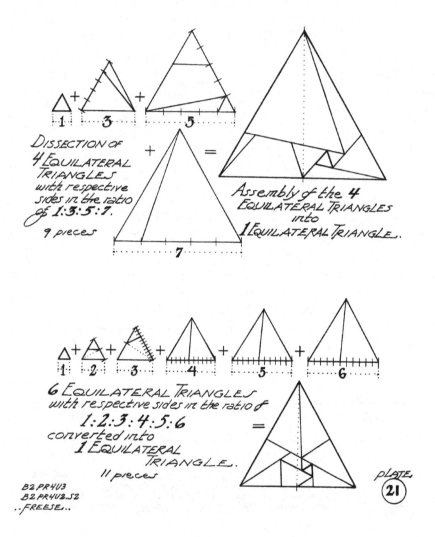

Figure M2.2.   Ernest Freese's Plate 21.

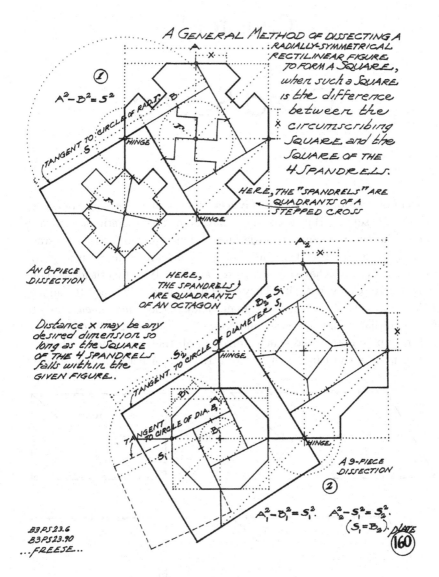

Figure M2.3.   Ernest Freese's Plate 160.

Chapter 5

Quad Squad

All you polygons out there probably feel that if you've got four sides, then you gotta be square. But hey, you can have four sides and still do your own thing. Who says you can't bend the rules and measure up in your own way? In this episode, we'll see three hip polygons that are really with it, if you know what I mean. They've figured out their own style of rectitude without being square. Can you dig it? They rebel from those conversion techniques of Chapters 3 and 4—too straight, man! They're the *Quad Squad,* ready to show quadrilaterals, trapezoids, and rectangles some real polygon power.

First up is a trapezoid that's hinged in all the right places. She sure is a beauty, as you can see in Figure 5.1. Take in her two long straight

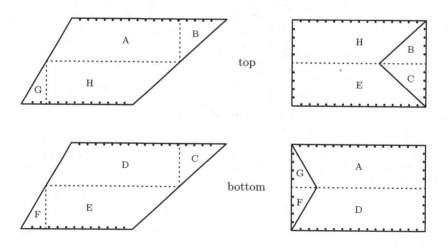

Figure 5.1. Folding a trapezoid to a rectangle of the same height. (C)

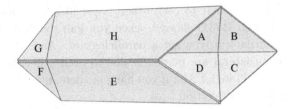

Figure 5.2. View of a tube-cyclic dissection: trapezoid to a rectangle.

bases, parallel to each other, between sides set at cool, rakish angles. And when you least expect it, she flips to a rectangle with no change in height! She scores with that special technique unless perpendiculars dropped from a midpoint of a side to both bases fall outside the endpoints of each base. Her 8-piece dissection, with a projected view in Figure 5.2, is tube-cyclic and has two flat-cycles—groovy! There's another way to hinge her pieces, which you'll have to get the lowdown on yourself.

Puzzle 5.1. *Find another way to hinge the pieces in Figure 5.1 to produce a piano-hinged dissection of a trapezoid to a rectangle of the same height.*

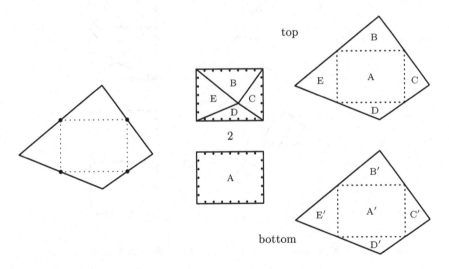

Figure 5.3. Quadrilateral with perpendicular diagonals to two rectangles.

Next up is a quadrilateral with attitude. He may come from a large family and have sides at wild angles, but he's got his own special bag: His diagonals are perpendicular to each other. Now like this ancient French dude Pierre Varignon (1731) showed, when you join the midpoints of the sides of any quadrilateral, you get a parallelogram. And when the diagonals of the quadrilateral are perpendicular to each other, the Varignon parallelogram is a rectangle. And if you hang in there, man, you'll see that this leads to a folding dissection.

Check out the one-level-thick convex quadrilateral on the left of Figure 5.3, which folds along the sides of the rectangle to give a second rectangle. For the corresponding two-level-thick quadrilateral, each of the bottom and top levels does this, so that we get a 10-piece folding dissection, as on the right in Figure 5.3. Just as Figure 1.8 proves that the sum of the angles in a triangle equals 180°, this figure proves that the sum of the angles of this type of quadrilateral equals 360°. That's right, man—this is a generalization of the triangle-to-two-rectangles dissection in Figure 1.8. In that case, the triangle is a degenerate version of a quadrilateral with perpendicular diagonals in which the bottom angle has gone straight (i.e., become 180°), and thus piece D in Figure 5.3 has flown the coop (i.e., vanished).

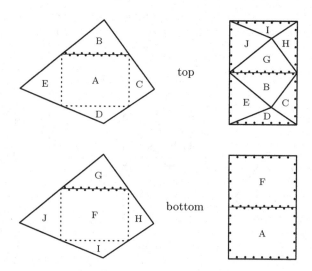

Figure 5.4. Quadrilateral with perpendicular diagonals to one rectangle. (C)

Okay, okay. What's bugging you, man? You uptight just because Donald Bruyr (1963) already cut a quadrilateral into five pieces, one of which was this Varignon parallelogram, and the other four formed a swing-hinged model of the same? Bruyr probably also noticed that you could cut the

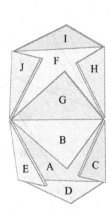

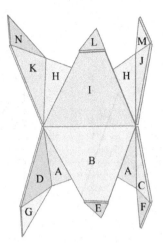

Figure 5.5. Quad. to a rectangle.   Figure 5.6. Nonconvex quad. to rect.

quadrilateral into four pieces that form one parallelogram with the same angles and base as the Varignon parallelogram, but twice as high. Lay off me, or I won't let you eyeball a simple piano-hinged dissection of a convex quadrilateral with perpendicular diagonals into a single rectangle (Figure 5.4). And stay loose, since we'll need a piano-hinging that is at least similar to the leaf-cyclicly hinged one in Figure 5.5. This leaf-cyclic hinging connects pieces A, B, F, and G. Since the quadrilateral is still convex after rotating it 90°, it's easy as one-two-three to find a different piano-hinged dissection into a rectangle of height equal to the Varignon parallelogram and length twice as long. Solid!

What's that? You think we're headed for a bad scene if that quadrilateral is not convex. No problem, man. We'll find a brother with more pieces to handle the job. When one of the vertices of the quadrilateral falls in the interior of the rectangle, as on the left in Figure 5.7, we get a 14-piece piano-hinged dissection, as on the right. The pieces are what we get when we superpose a grid of rectangles over the quadrilateral and cut along the grid lines. We'll have to can the dissection in Figure 5.7 if the vertex whose angle is greater than 180° falls outside of the rectangle. But a corresponding dissection, based on the grid forcing more pieces, does the trick.

And how 'bout a simple piano-hinged dissection of a nonconvex quadrilateral with perpendicular diagonals to a single rectangle? Would I string

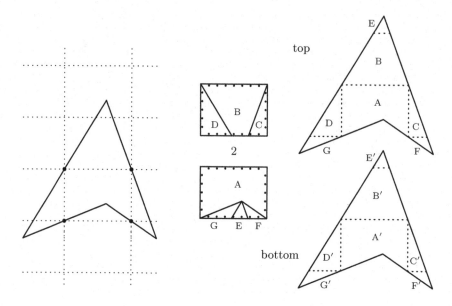

Figure 5.7. Nonconvex quadrilateral to two rectangles.

you along? Well maybe, but this time I've got a 14-piece dissection in Figure 5.8. Again, we seem to need a hinging at least close to being leaf-cyclic, and there's one strutting its stuff in Figure 5.6. And not to worry, there's also a rectangle whose base equals the width of the quadrilateral.

Now that we've seen how to deal with a nonconvex quadrilateral, maybe we are ready to deal with a trapezoid that bends the rules a bit further too.

Puzzle 5.2. *Find a 10-piece folding dissection of a trapezoid to a rectangle of the same height in the case that only one of the two sides of the trapezoid can drop a perpendicular that falls within the endpoints of a base.*

So far we've seen that hip trapezoid and the quadrilateral with attitude. The third member of our unconventional trio is already a rectangle, but one with identity problems. He wants to be a different rectangle. To pull this off, he's put the snatch on a simple technique from the domain of unhinged dissections, namely the *plain-strip*, or *P-strip*, technique. The setup is to unfold a two-level-thick rectangle into a one-level-thick rectangular strip element and to do the same with the other two-level-thick rectangle. Then, form an infinite strip from elements of one of the rectangles, and form a second strip from elements of the other rectangle. Then, "crosspose" these

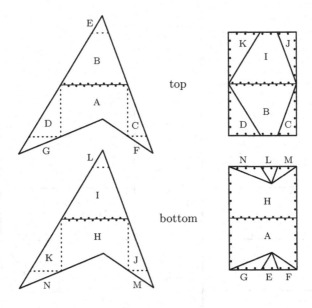

Figure 5.8. Nonconvex quadrilateral to one rectangle. (C)

strips so that the two boundaries of one strip cross one of the boundaries of the other strip at points that are at a distance equal to the length of the elements of the other rectangle. You can check out two examples of crossposing strips in Figure 5.9. The crossposition on the left in Figure 5.9 produces the 8-piece folding dissection in Figure 5.10. And strictly on the level, man: Pieces B, C, D, and E are flat-cyclicly hinged.

We can also use the *twinned-strip*, or *T-strip*, technique. Applied to rectangles, this technique is a restricted version of the P-strip techniques. When we crosspose the T-strips, we gotta overlay the points of central symmetry in each and force the boundaries of each strip to go through the neighboring points of symmetry of the other strip. Small dots identify these points of symmetry. At its best, the technique ends up in a dead heat with the eight pieces produced by plain strips. This happens when the symmetry points fall advantageously as on the right in Figure 5.9. However, when applied to more interesting examples, such as five equal squares to two in Figure 10.6, this method comes out on top.

Hovering over our young trio is a tough but ambivalent polygon. Is that a square that flips into a rectangle, or a rectangle that reverts to being square? Will the P-strip deliver for any pair of equal-area rectangles, no matter how long and thin one is in relation to the other? Let l and

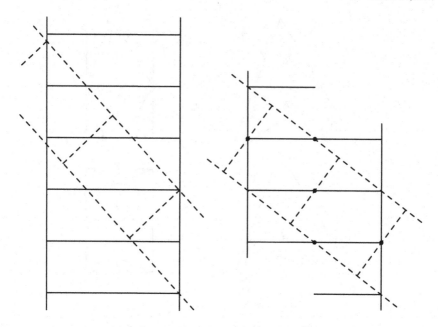

Figure 5.9. Crosspositions for rectangle to rectangle.

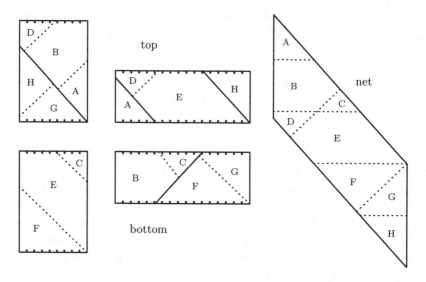

Figure 5.10. Folding dissection of one rectangle to another.

w be the length and width of the first rectangle, and αl and w/α be the length and width of the second. For α in the range $1 < \alpha < 2$, the strip method bombs when compared with a method that we shall see in the next chapter. But it lucks out for larger α, with the number of pieces being approximately $2\alpha + 4$. We can get a positive ID on the number of pieces needed in a fashion similar to, but more involved than, what Henry Taylor (1905) did for unhinged dissections. Let $\beta = \lceil \sqrt{\alpha^2 - 1} \rceil$. The strip method then uses $2\beta + 4$ pieces if $\alpha^2 \leq (\beta^2 + \sqrt{\beta^4 - 4})/2$ or $\alpha^2 = \beta^2 + 1$, and $2\beta + 5$ pieces otherwise.

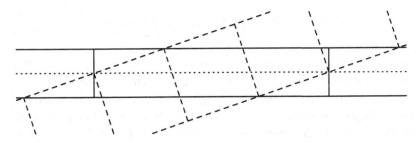

Figure 5.11. Crossposition for a (10×1)-rectangle to square.

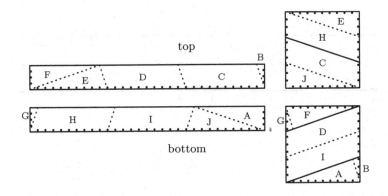

Figure 5.12. Folding dissection of a (10×1)-rectangle to a square.

The case of $\alpha^2 = \beta^2 + 1$ corresponds to the case when $\sqrt{\alpha^2 - 1}$ is a natural number, which means that we are converting an $(\alpha^2 \times 1)$-rectangle to a square. When $\alpha = \sqrt{5}$, then $\beta = 2$, and there is an 8-piece dissection of a (5×1)-rectangle to a square (Figures 12.1 and 12.2). When $\alpha = \sqrt{10}$, then $\beta = 3$, and there is a 10-piece dissection of a (10×1)-rectangle to

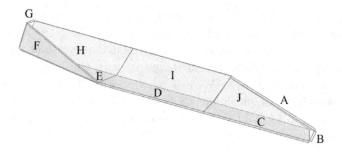

Figure 5.13.   View of a (10 × 1)-rectangle to a square.

a square. We see the crossposition for this dissection in Figure 5.11. The resulting folding dissection (Figure 5.12) is all right. A cool feature of it is that there are four sets of cyclic hinges: Pieces A, B, C, and J are cap-cyclicly hinged, as are pieces E, F, G, and H. In addition, pieces D, E, H, and I are flat-cyclicly hinged, as are pieces C, D, I, and J. These cycles will force the boat-like surface in Figure 5.13 as we fold out and down the pieces on the top level of the (10 × 1)-rectangle.

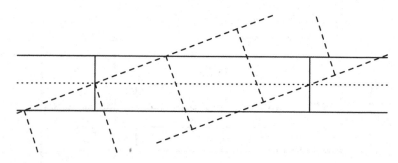

Figure 5.14.   Crossposition for an (8 × 1)-rectangle to square.

When $\alpha = \sqrt{8}$, then $\beta = 3$, and there is a 10-piece dissection of an (8 × 1)-rectangle to a square. The crossposition for this dissection appears in Figure 5.14, and the 10-piece dissection is in Figure 5.15. What a bum rap! We lose one of the cap-cyclic hingings, because we can no longer hinge pieces A and J. Sizing up the two crosspositions, we see that there is a range of α between $\sqrt{8}$ and $\sqrt{10}$ for which there will be one more crossing of lines in the crossposition, producing a total of eleven pieces. In that case, piece B would be replaced by two pieces, one on each level. This illustrates why

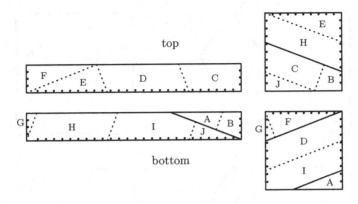

Figure 5.15. Folding dissection of an (8 × 1)-rectangle to a square.

we have the condition that $\alpha^2 \leq (\beta^2 + \sqrt{\beta^4 - 4})/2$ or $\alpha^2 = \beta^2 + 1$ in order to get the bound of $2\beta + 4$.

This quad squad takes its mission seriously, not hesitating to go undercover to protect other like-minded polygons of few sides from the many-sided polygons who might prey on them. With a determination mixed with a new groovy patriotism, they can even transform a flag. Henry Dudeney (*Dispatch*, 1903a) gave a novel adaptation of the step technique. In the "Shamrock Flag" puzzle, he dissected a flag shape into a square, using a piece formed by rotating a triangle about a point (Figure 5.16). To disallow the rather simple solution of cutting the triangle and filling the appropriate hole with it, Dudeney added a condition to the puzzle that effectively ruled out that cut. Dudeney's technique works for any triangle whose relevant corner divides evenly into 90°, giving the number of triangles in the rotating piece.

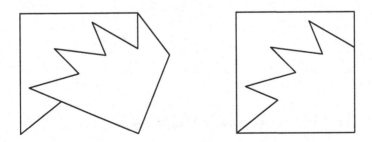

Figure 5.16. Dudeney's unhingeable dissection of a "flag" to a square.

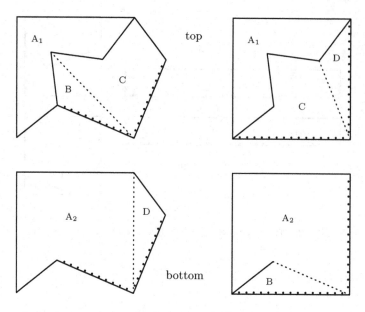

Figure 5.17.   Rounded piano-hinged "flag" to square.

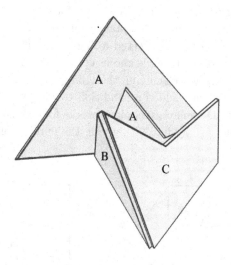

Figure 5.18.   Perspective view of the assemblage for a "flag" to square.

When we switch to piano-hinged dissections, Dudeney's rotating step suggests a nice solution for the flag puzzle (Figure 5.17). The crucial angle times an even number gives us 90°. In my 4-piece solution, pieces B and D are copies of the triangle, and piece C is the union of rotations and reflections of this triangle. Piece A is the remaining material. The four pieces are cap-cyclicly hinged, but this piano-hinged dissection is rounded. Starting from the "flag" and beginning to fold pieces B and C down and to the left gives us the perspective view in Figure 5.18.

Our daring, nonstandard polygons can even serve as role models for others. We can piano-hinge the dissection of one trapezoid to another if the trapezoids are the same height, their bases are long enough relative to the height, and the angles have a certain relationship. In particular, there must be a whole number greater than 1 that, when multiplied times the difference of the left obtuse angle of one trapezoid minus the left acute angle of the other, gives 90°, and similarly for the obtuse and acute angles on the right. We then get a 7-piece piano-hinged dissection like the one in Figure 5.19. Our example has a difference of 30° between the obtuse angle on the right and the resulting acute angle on the right, and a difference of 45° between the obtuse angle on the left and the resulting acute angle on the left. Note that piece F is the result of reflecting piece E across their common boundary, then reflecting piece E again, and so on. We may thus imagine piece F divided up into five copies of piece E that alternate between being flipped over and not flipped over. Similarly, we can divide up piece B into three copies of piece C.

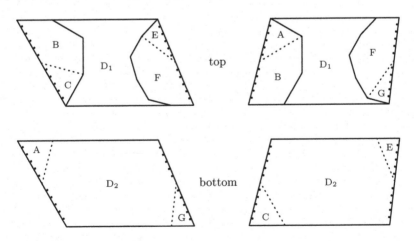

Figure 5.19. Piano-hinged trapezoid to another of the same height.

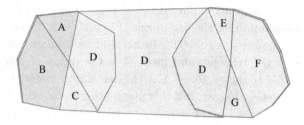

Figure 5.20. Projected view for a trapezoid to a trapezoid.

Now isn't the assemblage in Figure 5.20 a smooth way to make our exit? We have used the previous technique twice in this dissection, so that we have two flat-cyclic hingings. We rely on a corresponding 3-piece swing-hinged dissection, suggested by the bottom level if we ignore the labels of the pieces. When the trapezoids do not satisfy the required conditions, we may still be able to produce a 10-piece piano-hinged dissection by converting this corresponding 3-piece dissection, using the techniques of Chapters 3 and 4. But hey man, don't flip your lid! We wouldn't really fink on those trapezoids—that's not our bag. And besides, time's up for this show!

Chapter 6

Thinking Inside the Box

In the last chapter, we made the scene with trapezoids, quadrilaterals, and rectangles. In this chapter we will focus exclusively on rectangles, so we may naturally find ourselves "thinking inside the box," even though that is not the current fashion. We have already watched as strip techniques, camouflaged in the height of cool, have flipped one rectangle to another. Yet these techniques do not always produce the best results. Thus, we will study an even more elegant technique for transforming one rectangle to another, which we can use whenever the relative dimensions are within a factor of 2 of each other. With our thinking caps in place, we will then go on to extend our gift-wrapping technique in several ways.

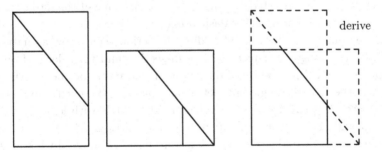

Figure 6.1. P-slide: Unhinged dissection of one rectangle to another.

In the nonfolding world, the *P-slide* is a fundamental technique that converts one rectangle to another. The "P" stands for parallelogram, on which it also works. The earliest documented appearance of it is in a dissection by Philip Kelland (1855). As we can see in Figure 6.1, the technique uses just three pieces. Its derivation (on the right in the figure)

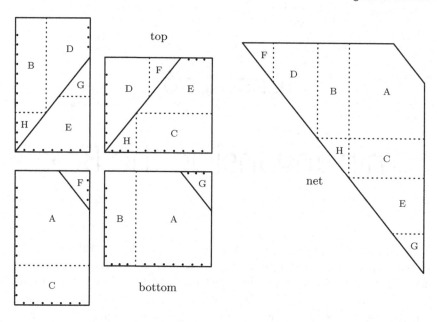

Figure 6.2. Folding dissection of one rectangle to another. (C)

is simple: Just overlay one rectangle on top of the other so that they share a corner. Then draw a line between the two corners that are furthest apart. We get one of the dissected rectangles by erasing some of the dashed lines and the other by erasing other dashed lines.

Once conceived, the related folding dissection seems just as natural. We will first design an 8-piece folding dissection and then derive from it a 7-piece rounded piano-hinged dissection. Let l and w be the length and width of the first rectangle, and let αl and w/α be the length and width of the second, where $1 < \alpha < 2$. Take a right triangle with legs of lengths $(\alpha + 1)l$ and $(1 + 1/\alpha)w$, and cut from it a small right triangle, with legs of lengths $(\alpha - 1)l$ and $(1 - 1/\alpha)w$. This produces a trapezoid whose bases are the hypotenuses of the two right triangles. Parallel to the leg of length $(\alpha + 1)l$, make cuts at distances of w/α, w, and $2w/\alpha$. Parallel to the leg of length $(1 + 1/\alpha)w$, make cuts at distances of l, αl, and $2l$.

We see an example in Figure 6.2, using the same pair of rectangles as in Figure 6.1. Pieces B and D together form the pentagonal piece in Figure 6.1, as do pieces C and E. Pieces D and F together form the large triangle in Figure 6.1, as do pieces E and G. Piece H corresponds to the small triangle in Figure 6.1.

When we fold to produce one of the two rectangles, exactly one of pieces B and C is on the same level with piece A. If piece C is on the same level with A, then pieces B, D, E, G, and H are on the other level, with D extending over to the righthand side of piece A and E below piece D. Piece F then folds around to fill in the small right-triangular notch in piece A. If piece B is on the same level with A, then pieces C, D, E, F, and H are on the other level, with E extending up to the top side of piece A and D to the left of E. Piece G then folds around to fill in the small right-triangular notch in piece A.

We can think of this technique as *gift wrapping a rectangle*. A notable property of the technique is that the sides of one rectangle are parallel to the sides of the other. In addition, the pieces A, B, C, and H are flat-cyclicly hinged, as we can see in the perspective view (Figure 6.3).

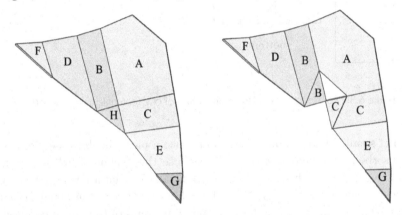

Figure 6.3.   Rectangle to rectangle.   Figure 6.4.   Holey assemblage!

There is a nifty way to save a piece in this case, if we are broad-minded enough to accept a rounded piano-hinged dissection. First delete piece H. Then add a corresponding triangle as the other level of piece B. This forces us to cut a corresponding triangle out of the lower left corner of piece A. To compensate, we then add a corresponding triangle as the other level of piece C. No matter which rectangle we form, one of the new triangles will fill the hole vacated by the former piece H, and the other will fill the hole next to piece A. Figure 6.4 gives a perspective view of the 7-piece assemblage, preceding the dissection itself in Figure 6.5. What was previously the net appears in modified form on the right in Figure 6.5, with the shaded triangles in pieces B and C indicating where the triangles would be attached.

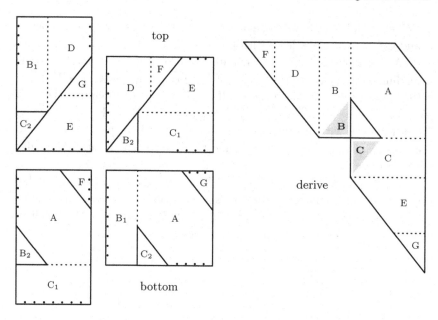

Figure 6.5. Rounded piano-hinged dissection of one rectangle to another. (W)

Let's call this technique *absorption*, since a piece, in this case piece H, is absorbed into another piece. We can use this technique whenever we have four pieces, in this case pieces A, B, C, and H, that are vertex-cyclicly hinged, meet at equal angles, and have hinges between the nonabsorbed pieces that are longer than the corresponding edges of the absorbed piece. The latter condition rules out performing absorption on piece D in Figure 2.6. Let's use the term *R-flap* for the technique that transforms one rectangle to another as in Figure 6.5.

It takes no great brainstorm to extend the gift-wrapping technique to handle a conversion factor of $\alpha > 2$. Again, take a right triangle with legs of lengths $(\alpha + 1)l$ and $(1 + 1/\alpha)w$, and cut a small right triangle, with legs of lengths $(\alpha - 1)l$ and $(1 - 1/\alpha)w$. Just make all possible cuts parallel to the leg of length $(\alpha + 1)l$ at distances that are whole number multiples of w/α and w, and all possible cuts parallel to the leg of length $(1 + 1/\alpha)w$ at distances that are whole number multiples of l and αl. This will give a folding dissection that contains $4\lceil \alpha \rceil$ pieces. Our example in Figure 6.6 satisfies $2 < \alpha < 3$, which means it has 12 pieces.

Readers may have the presence of mind to observe that we could just have rotated the second rectangle around to get an α that is less than 2.

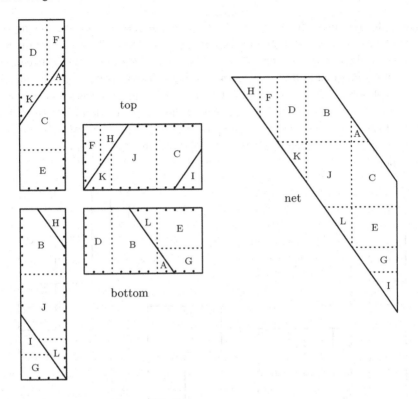

Figure 6.6. Folding dissection of one rectangle to another, with $2 < \alpha < 3$.

However, a few moments' reflection will convince you that such a rotation will not always be advantageous. If we divide the length of each rectangle by 10, then the techniques as presented would still produce a valid 12-piece piano-hinged dissection. Thus, I will illustrate the techniques with rectangles of a reasonable size for presentation, even as we understand that we could radically shrink one dimension with respect to the other.

The absorption technique also saves pieces when we apply it to examples in which $\alpha > 2$. In particular, we can find a rounded piano-hinged dissection with $3\lceil \alpha \rceil + 1$ pieces. You shouldn't need to rack your brain on the following puzzle.

Puzzle 6.1. *Find a 10-piece rounded piano-hinged dissection for the rectangles in Figure 6.6.*

In the Renaissance, people were already thinking up a storm while converting one rectangle to another. Girolamo Cardano (1663) converted a

$(3a \times 4b)$-rectangle to a $(4a \times 3b)$-rectangle in just two pieces, using a zigzag cut. He also converted a $(2a \times 3b)$-rectangle to a $(3a \times 2b)$-rectangle. This technique may well have been known earlier than Cardano, because Leonardo da Vinci (1452–1519) seems to have described it briefly in one of his notebooks (*Codex Atlanticus*). Leonardo sketched what is suggestive of the same technique, which has been called the "step technique." In my first book I showed how to derive the step technique from the P-slide technique, by swapping small triangles across a diagonal cut line, from one piece to another.

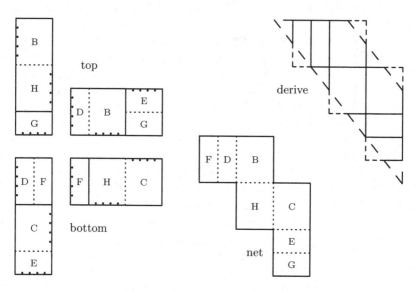

Figure 6.7. $(2a \times 5b)$-rectangle to a $(5a \times 2b)$-rectangle and derivation. **(C)**

In Figure 6.2 we saw an 8-piece folding dissection of an $(l \times w)$-rectangle into an $(\alpha l \times w/\alpha)$-rectangle that works whenever $1 < \alpha < 2$. Is it possible that for certain rational values of α, there are dissections with fewer than eight pieces? Can such dissections be produced by converting a diagonal cut to a staircase-shaped cut? If we switch on our ruminators, what can we dream up?

Let's start with the dissection of a $(2a \times 5b)$-rectangle to a $(5a \times 2b)$-rectangle that we can produce with our gift wrapping technique. I show a net for this 12-piece dissection on the upper right in Figure 6.7, using dashed and solid edges. If we swap small triangles, by deleting the longer dashed edges and inserting the shorter dashed edges, we get the net in the

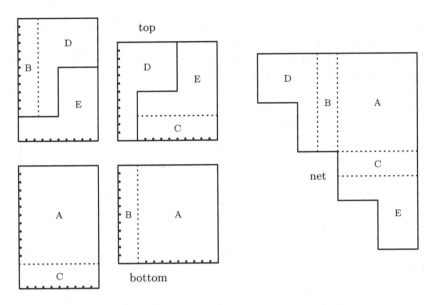

Figure 6.8. Folding a $(4a \times 5b)$-rectangle to a $(5a \times 4b)$-rectangle-rectangle. (C)

lower middle of the figure. This leads to the 7-piece dissection on the left. We will call this technique a *flap-step conversion*.

For α in the range $1 < \alpha < 2$, I have found two classes of rectangles for which there are folding dissections with fewer than seven pieces. When $\alpha = (2k+1)/(2k)$, for any whole number k, there is a 5-piece folding dissection that converts a rectangle to another whose length is increased by a factor of $(2k+1)/(2k)$ and whose width is decreased by the factor $2k/(2k+1)$. Therefore, for any positive lengths a and b, we can convert a $(2ka \times (2k+1)b)$-rectangle to a $((2k+1)a \times 2kb)$-rectangle. The dissection is nettable. We see an example in Figure 6.8, where $k = 2$.

Piece A is a $(2ka \times 2kb)$-rectangle, piece B is an $(a \times 2kb)$-rectangle, and piece C is a $(2ka \times b)$-rectangle. We figure out the shapes of pieces D and E by starting from the point that pieces B and C meet and fitting a staircase of $2k$ steps with a tread-length of $2a$ and a tread-height of $2b$. When we fold to produce one of the two rectangles, exactly one of pieces B and C is on the same level with piece A. Figure 6.10 gives a perspective view.

There is a nifty way to explain the dissection, in terms of reducing the puzzle of finding a piano-hinged dissection to a puzzle of finding an unhinged dissection. We can produce the piano-hinged dissection of a

$(2ka \times (2k+1)b)$-rectangle to a $((2k+1)a \times 2kb)$-rectangle by cutting piece A and then finding a simple unhinged dissection of a $(2ka \times 2(k+1)b)$-rectangle to a $(2(k+1)a \times 2kb)$-rectangle. Afterwards, we need to cut pieces B and C from the two pieces of that unhinged dissection.

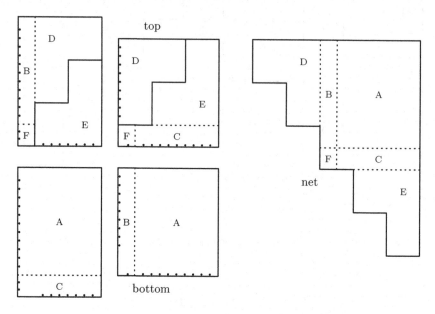

Figure 6.9. Folding a $(5a \times 6b)$-rectangle to a $(6a \times 5b)$-rectangle.

An analogous approach does almost as well when $\alpha = (2k+2)/(2k+1)$, for whole number k. In that case there is a 6-piece folding dissection. The dissection is nettable, with the net produced in a similar fashion. The example for $k = 2$ is in Figure 6.9. Piece A is a $((2k+1)a \times (2k+1)b)$-rectangle, piece B is an $(a \times (2k+1)b)$-rectangle, piece C is a $((2k+1)a \times b)$-rectangle, and piece F is an $(a \times b)$-rectangle. We can determine the shapes of pieces D and E by starting from the lower left corner of piece F in the net and fitting a staircase of $2k+1$ steps with a tread-length of $2a$ and a tread-height of $2b$. Note that pieces A, B, C, and F are flat-cyclicly hinged.

Wait a moment! Let's pause, collect our thoughts, and contemplate the similarity of Figures 6.2 and 6.9. If we forgo folding for rounded piano-hinging and use our absorption technique, we can do just as well. First, delete piece F. Then, add a corresponding rectangle as a second level of piece B. This forces us to cut a corresponding rectangle out of the lower left corner of piece A. To compensate, we then add a corresponding rectangle

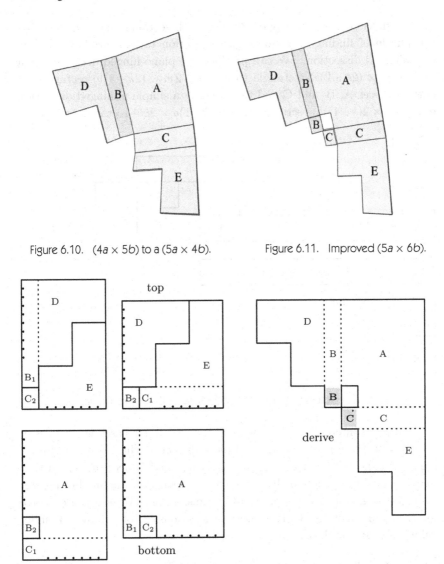

Figure 6.10.   (4a × 5b) to a (5a × 4b).   Figure 6.11.   Improved (5a × 6b).

Figure 6.12.   Rounded (5a × 6b)-rectangle to a (6a × 5b)-rectangle. (C)

to be the other level of piece C. No matter which rectangle we form, one of the new rectangles will fill the hole vacated by the former piece F, and the other will fill the hole in piece A. The resulting 5-piece dissection is in Figure 6.12. We see a perspective view in Figure 6.11.

Again, there is a nifty way to explain the dissection in terms of reducing the puzzle of finding a piano-hinged dissection to the puzzle of finding an unhinged dissection. We can produce the piano-hinged dissection of a $((2k+1)a \times (2k+2)b)$-rectangle to a $((2k+2)a \times (2k+1)b)$-rectangle by cutting pieces A, B, and C and then finding a simple unhinged dissection of a $(2ka \times 2(k+1))b)$-rectangle to a $(2(k+1)a \times 2kb)$-rectangle.

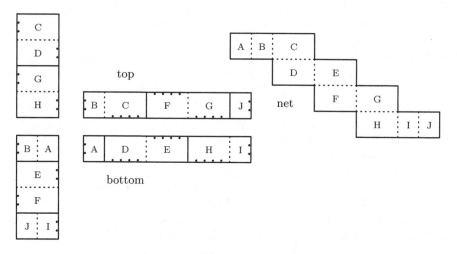

Figure 6.13.   Folding an $(a \times 4b)$-rectangle to a $(4a \times b)$-rectangle. (C)

There are interesting piano-hinged dissections for cases when $\alpha > 2$. (For $\alpha = 2$, there is of course the 4-piece dissection that we have already seen in Figure 2.9.) For $\alpha = 3$, there is a rounded 7-piece dissection that makes a nice puzzle for readers to ponder. For α a whole number greater than 3, there is a $(2\alpha+2)$-piece folding dissection. We see the example for $\alpha = 4$ in Figure 6.13. The folding dissections for other whole number values of α are similar.

Puzzle 6.2. *Find a 7-piece piano-hinged dissection of an $(l \times w)$-rectangle to a $(3l \times w/3)$-rectangle.*

Having spent this whole chapter in the box, readers may be starting to feel a bit claustrophobic. As we move on to the next chapter, we're not yet ready to abandon the box, but we will cut a small rectangular hole in its side to let in some fresh air and light. Now isn't that thoughtful?

Chapter 7

Hole in One

In the world of golf, a hole-in-one is an extraordinarily rare and highly prized occurrence: You smack a small ball a couple hundred yards, and it winds up rolling into the "hole," a cup that is only a few inches in diameter. For this you need not only tremendous skill but also, and to a much greater extent, remarkable luck. In the world of dissections, things aren't quite as difficult, but dealing with a hole is still a tricky proposition. Several dissectionists have scored with a hole-in-one, at least when the object containing the hole was a rectangle, the hole itself was a centered rectangle, the resulting object was a square, and the dissection was not hinged. Can we match their accomplishments if we aim for piano-hinged dissections?

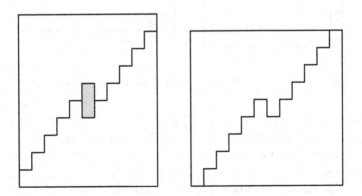

Figure 7.1. Dudeney's unhingeable (33 × 40)-rectangle with a hole to a square.

In the "early days" of this sport, Henry Dudeney (*Strand*, 1926a) gave a 2-piece unhingeable dissection of a (33 × 40)-rectangle with a (3 × 8)-rectangular hole to a 36-square. His solution (Figure 7.1), uses a clever zigzag cut. Such an approach is initially striking, until we realize that we can work forward: We deduce that the first cut must be a certain zig, which forces the second cut to be a certain zag, and so on, until we have formed the needed compound cut.

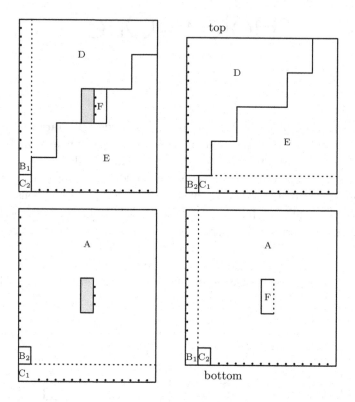

Figure 7.2.   Rounded (33 × 40)-rectangle with a 3 × 8 hole to a square.

However, this approach does not give much of a hint of how to find a corresponding piano-hinged dissection. Fortuitously, there is a 6-piece rounded piano-hinged dissection, as we see in Figure 7.2. We see a perspective view of the piano-hinged assemblage in Figure 7.3. First, we cut a piece F from the top level that will fold around to exactly fill the (3 × 8)-rectangular hole on the bottom level. Second, we cut piece A as large as possible so that it will occupy the bottom level of the rectangle and the bottom level of the square. We use the trick in Figures 6.8 and 6.12 to fill out the bottom level of each by pieces B and C that will switch between the two levels when going from the holey rectangle to the square. Third, we cut what remains on the top level into pieces D and E that fit together in either one fashion or the other. Then we break out the champagne—a hole-in-one!

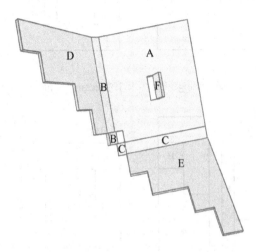

Figure 7.3. Assemblage for (33 × 40)-rectangle with 3 × 8 hole.

Does this dissection fit into any pattern? The answer is yes, although it is not so easy to spot it at first. Since all cuts are parallel to the sides of the rectangle, we can get an equivalent dissection by multiplying or dividing each horizontal dimension by a fixed amount. The same idea works for the vertical dimensions. Dividing the horizontal dimensions by 3 and the vertical dimensions by 4, we see that Dudeney's original puzzle is equivalent to dissecting an (11 × 10)-rectangle with a (1 × 2)-rectangular hole to a (12 × 9)-rectangle.

It is not difficult to find a generalization to other rectangles with a (1 × 2)-rectangular hole that have 2-piece unhinged dissections. Another example is a (9 × 8)-rectangle with a (1 × 2)-rectangular hole to a (10 × 7)-rectangle. Some of these examples have 6-piece dissections like the one in Figure 7.2, and some do not. Let's explore generalizations after we have had another round of practice.

Dudeney (1931) also gave a 2-piece dissection of a (9 × 12)-rectangle with a (1 × 8)-rectangular hole into a 10-square. What a gorgeous puzzle, with a

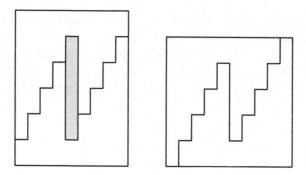

Figure 7.4. Dudeney's unhingeable (9 × 12)-rectangle with a hole to a square.

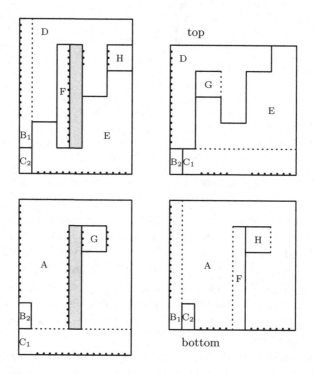

Figure 7.5. Rounded (9 × 12)-rectangle with a 1 × 8 hole to a square.

clever solution (Figure 7.4)! At first, the piano-hinged version of this puzzle seemed rather difficult to solve. After several muffed shots, I was able to scramble and find the 8-piece rounded piano-hinged dissection in Figure 7.5.

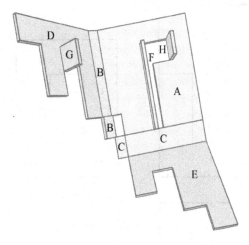

Figure 7.6. Assemblage for (9 × 12)-rectangle with 1 × 8 hole.

A perspective view (Figure 7.6) helps us visualize this dissection. Once again we see the trick of folding piece F from the top level of the rectangle to fill the rectangular hole on the bottom. We also see how pieces D and E will fit together in two different ways on the top level. However, to get them to fit together in this puzzle, it seems difficult to avoid using two extra pieces G and H to shift surplus area from right to left as we go from the rectangle to the square. Yes, we have achieved a hole-in-one, but with eight pieces, can we even save par?

Actually, we can do better with a related set of instances, so let's pick up our ball and move on to the next hole. If we consider cases in which the length of the hole is 4 units smaller than the length of its enclosing rectangle, we can often find a 6-piece rounded piano-hinged dissection. As an example, Figure 7.7 shows a dissection of a (7 × 10)-rectangle with a (1 × 6)-rectangular hole to a square. Again we draw inspiration from the flap-step dissection of one rectangle to another. We cut a piece F to fold over and exactly fill the (1 × 6)-rectangular hole on the bottom level. We then cut what remains on the top level to fit together in a lovely fashion. As a bonus, we find that we can hinge pieces A, B, D, and F together tube-cyclicly, as we see in a perspective view of the piano-hinged assemblage in Figure 7.8.

Although he gave no indication that he was aware of it, Dudeney's two examples come from the same broad class of 2-piece puzzles: Let r and v be

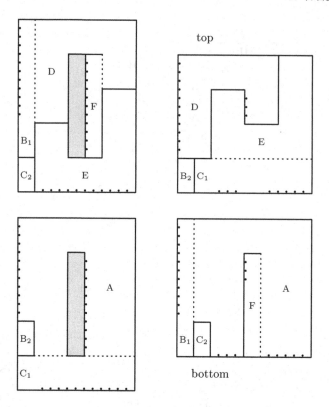

Figure 7.7. Rounded (7 × 10)-rectangle with a 1 × 6 hole to a square. **(W)**

natural numbers that will help specify horizontal and vertical expansion, respectively. Let m, p, and q be natural numbers, where $2m$ specifies the number of steps, each of length p, and q specifies the length of the hole, where $q < (m{+}1)p$. For any such values of these variables, there is a 2-piece dissection of an $(r(2m{+}1) \times v((2m{+}2)p - q))$-rectangle with a $(r \times vq)$-rectangular hole to a $(r(2m{+}2) \times v((2m{+}1)p - q))$-rectangle. For example, setting $r = 1$, $v = 1$, $m = 4$, $p = 2$, and $q = 8$ gives the (9 × 12)-rectangle with the (1 × 8)-rectangular hole to a square that we saw in Figure 7.4.

If we want the latter figure to be a square, as Dudeney did, we find the least common multiple s of the length and width of the resulting rectangle. We then multiply all horizontal dimensions by s divided by the length and multiply all vertical dimensions by s divided by the width. The resulting rectangle will be an s-square.

Remarkably, this generalization leads us to a piano-hinged dissection by a reduction to a 2-piece puzzle! Let's follow the general form of Figures 7.2

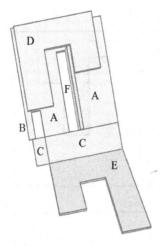

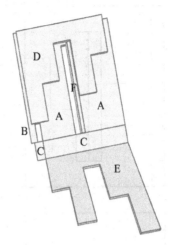

Figure 7.8. (7 × 10)-rectangle with hole.   Figure 7.9. (11 × 14) with hole.

and 7.7, using pieces A, B, C, and F as described earlier. We wind up needing to solve a 2-piece unhinged puzzle involving pieces D and E. As we reduce the piano-hinged puzzle to the 2-piece puzzle, the hole will become twice as wide, the rectangle with the hole will become narrower, and the rectangle without the hole will become shorter. Specifically, we will need to dissect a $(2mr \times v((2m+2)p - q))$-rectangle with a $(2r \times vq)$-rectangular hole to a $(r(2m+2) \times v(2mp - q))$-rectangle.

We divide each horizontal dimension by $2r$ and each vertical dimension by v to get the equivalent puzzle of an $(m \times ((2m+2)p - q))$-rectangle with a $(1 \times q)$-rectangular hole to an $((m+1) \times (2mp - q))$-rectangle. To get this into our expected form for 2-piece puzzles, we set $p' = 2p$. Our resulting puzzle is thus to dissect an $(m \times ((m+1)p' - q)$-rectangle with a $(1 \times q)$-rectangular hole to an $((m+1) \times (mp' - q))$-rectangle. And we know how to do this whenever m is odd! Note that the initial requirement of $q < (m+1)p$ implies that $q < m'p'$, where $m' = (m-1)/2$, which certainly implies $q < (m'+1)p'$. Thus, we can find a 6-piece rounded piano-hinged dissection whenever m is an odd number.

In addition, we uncover the condition that allows a cyclic hinging among pieces A, B, D, and F: Piece D must have a portion flush with piece F to the left of piece F on the top level of the rectangle with the hole in it. This happens whenever $q > p' = 2p$. Therefore, the condition for the cyclic hinging when m is odd is $q > 2p$. The dissection in Figure 7.7 satisfies this

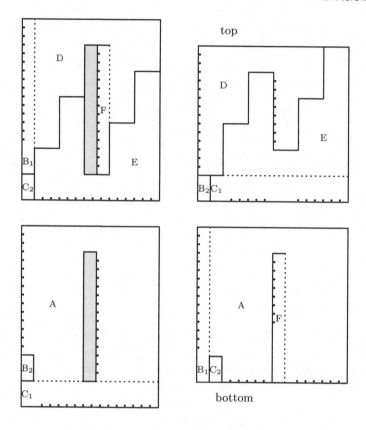

Figure 7.10. Rounded (11 × 14)-rectangle with a 1 × 10 hole to a square.

condition and thus has a cyclic hinging, while the dissection in Figure 7.2 does not.

As a further example, we find a 6-piece rounded piano-hinged dissection of an (11 × 14)-rectangle with a (1 × 10)-rectangular hole to a 12-square in Figure 7.10. Applying the general form, we see that $r = 1$, $v = 2$, $m = 5$, $p = 1$, and $q = 5$. Since $q > 2p$, we expect to see a cyclic hinging, which we do in fact have, again shown in perspective (Figure 7.9). The top levels of the two rectangles contain a 2-piece dissection of a (10 × 14)-rectangle with a (2 × 10)-rectangular hole to a (12 × 10)-rectangle. Do you see it?

Puzzle 7.1. *Find the corresponding 2-piece unhingeable dissection of a (11 × 14)-rectangle with a (1 × 10)-rectangular hole to a square.*

In the "modern era," Harry Langman (1962) posed the puzzle of finding a 2-piece dissection of a 5-square that has a 1-square hole in its middle into a (6 × 4)-rectangle. Langman, a mathematician on the faculty of colleges such as the Detroit Institute of Technology, Ohio Northern University, and Ball State Teachers College, posed this as an isolated puzzle. However, we should not be surprised that it fits our generalization in terms of r, v, m, p, and q. In particular, $r = 1$, $v = 1$, $m = 2$, $p = 1$, and $q = 1$.

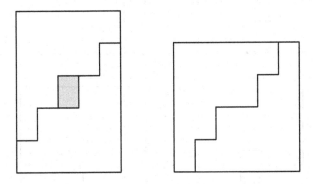

Figure 7.11.   Unhingeable (10 × 15)-rectangle with a 2 × 3 hole to a square.

It is easy to redesign Langman's puzzle so that it is a dissection of a rectangle with a rectangular hole into a square, as we have already discussed. Doing this gives us a (10 × 15)-rectangle with a (2 × 3)-rectangular hole to a 12-square. We see the corresponding 2-piece unhingeable dissection in Figure 7.11.

Our previous solution technique will not give us a piano-hinged dissection in this case, because m is an even number. Yet, we must be living right—our ball takes a lucky bounce—for we can find the 6-piece solution in Figure 7.12. Piece F folds down from the top level of the holey rectangle to fill in the hole in the bottom level of the square. A nifty feature is that we have a cyclic hinging involving pieces A, C, E, and F. Whenever m is even and $q = p$, the corresponding piano-hinged dissection is tube-cyclicly hinged! Notice that neither piece B nor piece C extends into both levels. This means that unlike the previous piano-hinged dissections in this chapter, this one needs no rounding, as we see in the perspective view (Figure 7.13).

What a terrific way to end our round! Let's amble up to the clubhouse, find a cozy corner in the 19th hole, and share with friends the tales of our success.

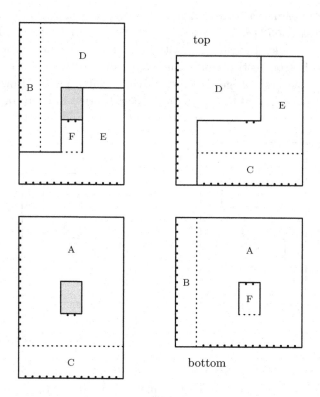

Figure 7.12. Rounded (10 × 15)-rectangle with a 2 × 3 hole to a square.

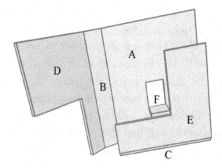

Figure 7.13. Assemblage for (10 × 15)-rectangle with 2 × 3 hole.

Folderol 1

A New Wrinkle
on an Old Problem

A problem that has become a staple in calculus textbooks is the *box problem*: Determine an open box of largest volume that we can form from a rectangular sheet by cutting squares out of the corners, folding up the sides, and then gluing or soldering the joints. Isaac Todhunter (1852) included it as an exercise in his calculus textbook a century and a half ago. The problem is a variation of a considerably older problem, which corresponds to the task of forming one quarter of an open box. That older problem was posed by the seventeenth-century French mathematician Pierre de Fermat and solved by the Dutch mathematician Frans van Schooten (1659).

The box problem entered the mathematical puzzle literature when it appeared in Henry Dudeney's puzzle column (*Dispatch*, 1903b). He later rephrased it for another puzzle column (*Cassell*, 1908a), adding a picturesque illustration by Paul Hardy, before he included the revised puzzle in his book *Amusements in Mathematics* (1917). The original version is delightfully quaint:

No. 525.–HOW TO MAKE CISTERNS.

Here is a little puzzle that will elucidate a point of considerable importance to cistern makers, ironmongers, plumbers, cardboard-box makers, and the public generally.

Our friend the cistern-maker has an interesting task before him. He has a large sheet of zinc, measuring eight feet by three feet, and he proposes to cut out square pieces from the four corners (all, of course, of the same size), then fold up the sides, join them with solder, and make a cistern.

So far, the work appears to be pretty obvious and easy. But the point that puzzles him is this: What is the exact size for the square pieces that he must cut out if the cistern is to contain the greatest possible quantity of water?

Call the feet inches, and take a piece of cardboard or paper eight inches long and three inches wide. By experimenting with this you will soon see that a great deal depends on the size of those squares. To get the greatest contents you have to avoid cutting those squares too small on the one hand and too large on the other. How are you going to get at the right dimensions? **I SHALL AWARD OUR WEEKLY HALF-GUINEA PRIZE.** for a correct answer. State the dimensions of the squares and try to find a rule that the intelligent working man may understand.

Figure F1.1.   Illustration from *Cassell's Magazine*.

The illustration from the later version of the puzzle appears in Figure F1.1. I invite readers to guess why the two men look grumpy. The solution to the original version of the puzzle appeared two weeks later (Dudeney, *Dispatch*, 1903c):

This was a little puzzle of a very practical and useful character. Given an oblong sheet of zinc, how should the workman cut out a square piece from each corner so that the four sides fold up and make a cistern that shall contain the largest possible quantity of water? The rule is simply this: (1) Deduct the product of the sides from the sum of their squares; (2) find the square root of the remainder; (3) deduct this square root from the sum of the sides; and (4) divide the remainder by 6. The result is the side of the little square pieces to be cut away.

Let us apply this rule to a sheet of zinc of the given dimensions, eight feet by three feet. (1) The sum of the squares of these two numbers is $64 + 9 = 73$, from which deduct $8 \times 3 = 24$, and we get 49. (2) The square root of 49 is 7. (3) Deduct 7 from $8 + 3$ and we have 4. (4) Now, if we divide four feet by 6, we get eight inches as the side of the square pieces.

This is the correct answer that we want. The intelligent working man is supposed in these days to know that a number multiplied by itself is a square, and that this number is called the "root" of such a square. Even if he does not know how to find the square root of any number, there are always table books available. I therefore think it best to give the exact method instead of one of the many approximations that have been suggested.

Try the rule in the cases of sheets measuring 8 feet by 5 feet, 16 by 6, 16 by 10, and 21 by 16, and you will find that the answers work out 1, $1\frac{1}{2}$, 2, and 3 respectively. Of course, it will not always come out exact (on account of that square root), but you can get it as near as you like with decimals.

The prize has been awarded to Mr. W. Robins, Wanstead Cottage, New Wanstead, Essex. Although the majority of competitors were considerably out in their calculations, 34 correct answers were received. Some of these came from persons who admittedly merely found by trial that it was "somewhere near eight inches;" and then ventured a guess that it was eight inches exactly. Others may have done the same, so there will be no honourable mention on this occasion.

The method that Dudeney describes is exactly the one that we can derive by calculus: Let x be the length of each square piece. Form the volume of the cistern in terms of x and the sides of the sheet, take the first derivative of the volume with respect to x, set the result to zero, and then solve using the quadratic formula. However, instead of discussing how to find the method, or equivalently, why the method is correct, Dudeney asserted that intelligent people ought to know what a square root is!

Figure F1.2 illustrates the method on a sheet of metal that is 3×4. In this case $x = (7 - \sqrt{13})/6 \approx .5657$. As directed, we cut out the four square corners in Figure F1.2. We then use rectangle A as the base of the cistern and fold rectangles B, C, D, and E up for the sides. The volume is $(35 + 13\sqrt{13})/27 \approx 3.0323$.

The box problem is a nice application of calculus and leads to a variety of interesting related problems. Dick Stanley (2001) and Wally Dodge and Steve Viktora (2002) observed that the optimal solution has an intriguing property, namely that for any shape of rectangle, the total area of the sides of the box will equal the area of the bottom. They also proved the same property for a corresponding approach applied to sheets of metal whose boundaries are polygons. James Duemmel (1989), Al Cuoco (2000), and Philip Hotchkiss (2002) characterized pairs of integral dimensions for which x is a rational number. Richard St. André (1983) and Kay Dundas (1984) identified variations in which the box is self-bracing.

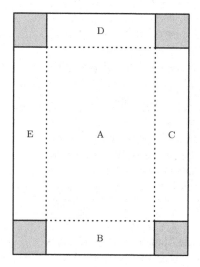

Figure F1.2.   Traditional cuts, folds.

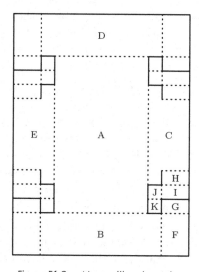

Figure F1.3.   New millennium plan.

Several people have suggested that the problem is a bit silly from a practical point of view. John Friedlander and John Wilker (1980), Kay Dundas (1984), and Donna Pirich (1996) pointed out that the corners are wasted. Friedlander and Wilker used this as an opportunity to apply the same technique recursively to the resulting squares, producing an infinite succession of boxes whose total volume is to be maximized.

Although it might be nice to have a large collection of boxes and at the same time to avoid waste, it might be nicer to have just one container to hold water. Who says that a cistern must be in the shape of a rectangular solid? Thus, I ask for a container of any shape, open on the top, formed from a rectangular sheet of metal, using tin snips and solder, that maximizes the volume. To retain the emphasis on folding from the original version of the problem, let's require that all material to be used in the container remain connected after the cutting and before the folding and soldering. To make matters simpler, let's consider here only containers whose sides are parallel to the sides of a cube and are of only single thickness.

We quickly discover that we can do better than soldering together the infinite number of boxes produced by Friedlander and Wilker. We need not form the sides that would be soldered together. In fact, for each small square of side x we can fashion a corresponding "bulge" in the cistern of volume $8x^3/27$. In Figure F1.3, I cut a sheet of the same shape as before, using solid edges to denote cuts and dotted edges to show folds. Suitably folded, this produces our cistern for the new millennium in Figure F1.4,

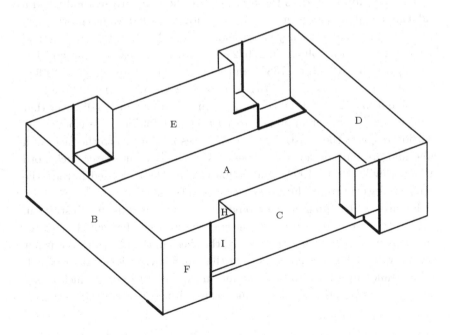

Figure F1.4. Perspective view of a new millennium cistern.

with the edges that we solder in bold and the folds shown with edges of normal thickness. We can easily see that panels J and K are small squares, and panels G, H, and I are twice as long as they are wide.

Since we no longer have small squares to discard, let's now use x to denote the height of the resulting cistern. Applying the standard calculus technique gives an optimizing value

$$x = (9/70)(a + b - \sqrt{a^2 + b^2 - 17ab/9}).$$

For $a = 3$ and $b = 4$, we get $x = (9/10)(1 - \sqrt{1/21}) \approx .7036$ and a volume of $(9/25)(9 + \sqrt{1/21}) \approx 3.3185$.

To compare with Friedlander and Wilker's construction, let's also consider the case that $a = b = 1$ (a square sheet of metal). They optimize with $x = \sin 10°$ and a volume of $(2/3)(1 - 2\sin 10°)\sin 10° \approx .0756$. By contrast, the textbook method, which Dudeney described, would have $x = 1/6$ and a volume of $2/27 \approx .0741$. My approach has $x = 3/14$ and a volume of $4/49 \approx .08163$, which beats Friedlander and Wilker by 8% and Dudeney by 10%.

The new approach does the best against the standard method when a and b are equal, and loses much of its advantage as these values grow apart. For the values that Dudeney chose, namely 3 and 8, the standard method has $x = 2/3$ and a volume of $200/27 \approx 7.4074$, whereas my method has $x = (3/70)(33 - \sqrt{249}) \approx .73801$ and a volume of approximately 7.81400. Even so, this is a gain of over 5%. Is it too late to collect that half-guinea?

Aside from any potential economic gain, it's rewarding to discover that there is a more clever approach to cutting and folding. Of course, my solution respects the constraint that the faces be parallel to the faces of a cube and are only of single thickness. Kay Dundas, and later Nick Lord (1990), Neville Reed (1992), and Kenzi Odani (2000), looked beyond the orthogonal world to produce even better solutions.

Regarding the apparent grumpiness of the two men in the illustration, perhaps they were so flustered by the square root that they cut the squares out of the wrong sheet of zinc. Clearly, the sheet in the illustration was not originally 8×3. Furthermore, they cut the squares with sidelength $x = a/4$, which cannot maximize the volume for any value of $b \geq a$. As luck would have it, $x = a/4$ is the smallest value for which this is regrettably true.

Chapter 8

Square Tactics

The headline shrieked its dire message:

SQUARES SHOW VULNERABILITY — FOLD UNDER PRESSURE!

Hearts raced and emotion ran rampant. Authorities agonized over whether a citizenry accustomed to unbending rules could remain calm while powerful techniques shook their very foundation and structure. The hysteria spread as later reports arrived:

TESSELLATIONS MAKE SHORT WORK OF SQUARES

HOUSE COLLAPSES AFTER FOLDING

R-FLAPS AT FAULT IN DEMISE OF SQUARES

How could these techniques menace the self-sufficiency and independence of all manner of squares? Would life ever return to the way it was before this frontal assault? Can anyone be safe when such square tactics are employed?

Postponing a response may leave us even worse off, so let's join in the action and try to help stem the math panic. One of the oldest known dissections (from the ninth century, C.E.) is Thābit ibn Qurra's 5-piece dissection of two unequal squares to one. Paul Mahlo (1908) and Percy MacMahon (1922) noted that the dissection results from superposing two tessellations, one consisting of pairs of the unequal squares and the other consisting of the large squares. (Given a plane figure, a *tessellation of the plane* is a covering of the plane with copies of the figure in a repeating pattern such that the copies do not overlap.) Furthermore, Thābit's dissection is swing-hingeable, and in my second book I gave a different superposition of tessellations from which we can derive the hinged dissection.

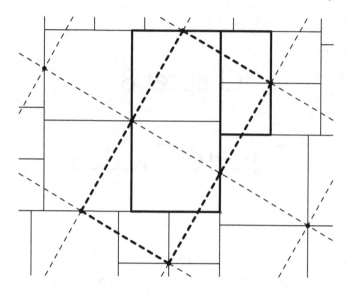

Figure 8.1. Superposition for folding two squares to one.

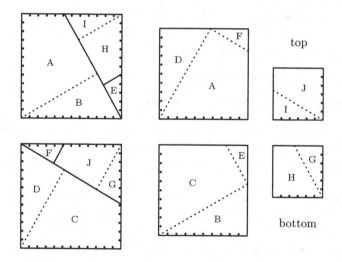

Figure 8.2. Folding dissection of two squares to one.

Here I lose all inhibition and show that we can adapt Thābit's dissection to make it folding. Indeed, I start with the two tessellations from the superposition that gave rise to a swing-hinged dissection and then push through

the folding dissection. We confront this superposition of the tessellations
in Figure 8.1, in which I outline in bold a pair of small squares and a pair
of medium-sized squares. Similarly, I isolate a pair of large squares with
thicker dashed lines. The actual 10-piece folding dissection lies flat before
us in Figure 8.2. We can see in Figure 8.3 that pieces A, B, C, and D are
cap-cyclicly hinged, as are pieces G, H, I, and J.

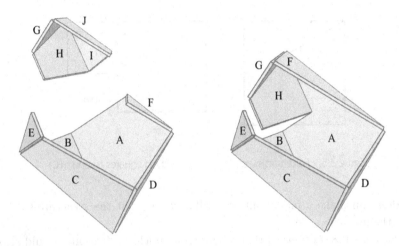

Figure 8.3. Two squares to one.   Figure 8.4. Two attached squares to one.

Do we gain any stability if the small and the medium-sized square are
attached together to form one piece? Not surprisingly, we find ourselves
cornered: The dissection in Figure 8.2 simplifies to give the 8-piece dissection in Figure 8.5. This is the same sort of simplification suggested by
a figure of Johann Sturm (1700) that illustrated the standard dissection
of two squares to one. As in the dissection of unattached squares, there
are two cap-cycles. You can study the collapse in progress in Figure 8.4.
With the same dimensions for the squares (unattached and attached) in
this figure and the last, it is easy to see the similarities between the two
dissections. To provide a better view of the assemblages, I have rotated
the images by 180°, so that we are viewing them from the rear.

There is one more suspicious feature of the folding dissection in Figure 8.5. A vertex of triangle E touches a vertex of trapezoid H in both
the large square and the two attached squares. I have marked this vertex
with a large dot in the figure. A natural question pops up: If we attach
these two vertices together by means of a universal joint, will we inhibit
the collapse of the large square to the two attached squares? Concerned

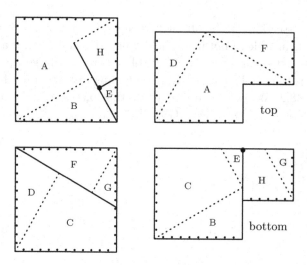

Figure 8.5. Folding dissection of two attached squares to one. (C)

readers can make a model out of cardboard and put the structure to the test themselves.

Brodie (1884) observed that a dissection much like Thābit's would also work for similar rectangles. Will folding also affect these rectangles?

Puzzle 8.1. *Find a 10-piece folding dissection for any two similar rectangles, assuming that the width and length of any one of the rectangles do not differ too much.*

There are several variants of the puzzle involving two attached squares to one. Sam Loyd (*Inquirer*, 1902a) introduced the puzzle of an isosceles right triangle attached with its hypotenuse flush against the side of a square, so that the triangle and the square share a vertex. Let's consider here a slightly simpler version that Loyd proposed (*Press*, 1900a), in which the hypotenuse of the isosceles right triangle is the same length as the side of the square to which it is attached. He presented the shape as the front of a doghouse, but let's simply call it a *house* here. Let's shake off thoughts of disaster as we view on the next page the lovely graphic for the "Young Carpenter's Puzzle," complete even to the detail of the carpenter's hats folded out of paper for the boys. Loyd gave a 3-piece dissection to a square, which I adapt to give the 6-piece folding dissection in Figure 8.6.

The Young Carpenter's Puzzle

PROPOSITION—Into how few pieces need the table top be cut to complete the dog house?

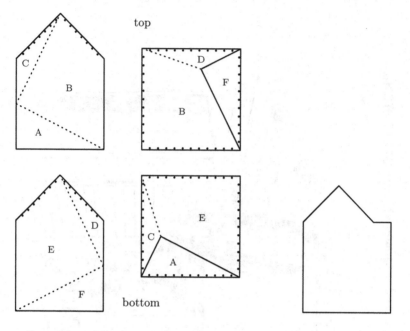

Figure 8.6. Folding house to square. **(C)**   Figure 8.7. Small-roofed.

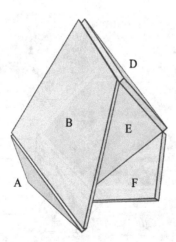

Figure 8.8. Perspective view of the assemblage for a house to a square.

The new dissection has a cap-cycle—where else but at the peak of the roof! I display a perspective view of the piano-hinged assemblage in Figure 8.8. The house's silhouette, formed by pieces A, B, and C (or alternately, pieces D, E, and F, behind them) has already started to crumble, and we see the silhouette of the square form with pieces B, D, and F folding up and around.

As for the more irregular shape (Figure 8.7) that Loyd first dissected, let's see if our housing inspectors can find an 8-piece folding dissection. Loyd called the shape a "remnant," though clearly it is just a *small-roofed house*.

Puzzle 8.2. *Find an 8-piece folding dissection of a small-roofed house to a square.*

As squares buckle all around us, beware of those opportunists who fold two squares to two different squares. In my first book, I gave a 6-piece dissection that is swing-hingeable and hinge-snug. Of course, we can convert that dissection to a 9-piece twist-hinged dissection and then apply our conversion technique from Chapter 4 to give a 16-piece folding dissection. However, danger is closer than you think: In my second book, I gave a 6-piece twist-hinged dissection for an extensive class that I called the Penta-large class. When we apply the conversion techniques, we get a 10-piece rounded piano-hinged dissection.

There is another whole class, orthogonal to the Penta-large class, that even without the use of conversion can suffer a similar fate. Take any three positive numbers (not necessarily whole numbers) a, b, and c such that $a^2 + b^2 = c^2$. It is always the case that $c^2 + c^2 = (a + b)^2 + (b - a)^2$. The new class consists of all such instances, and for each of these there is an 11-piece piano-hinged dissection. One level of a square that is of sidelength c will consist of four right triangles with legs of length a and b plus one level of a square of sidelength $|b - a|$. Thus, both levels of that square will fold out to one level of a square of sidelength $a + b$ plus one level of a square of sidelength $|b - a|$. Only the smallest square maintains its structural integrity in the 11-piece piano-hinged dissection in Figure 8.9. Using $z + w = 2b$ and $z - w = 2a$, we can restate the rule in terms of x, y, z, and w. How ironic that one piece (piece F) has a hole in it, through which to view the carnage in Figure 8.10!

Now the roof begins to cave in, as we spot another class that yields to similar techniques. For this class we see that $z = y\sqrt{2}$, allowing the y-square to fold out and produce one complete level of the z-square. The x-square then makes up the other level plus both levels of the w-square. We see the complete devastation in Figure 8.11. The x-square folds up to

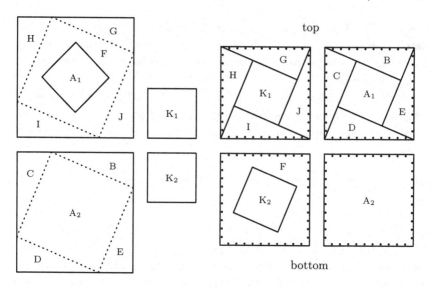

Figure 8.9. Squares for $x^2 + y^2 = z^2 + w^2$ where $(z+w)^2 + (z-w)^2 = 4x^2 = 4y^2$.

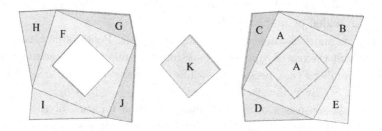

Figure 8.10. Assemblages for squares where $(z+w)^2 + (z-w)^2 = 4x^2 = 4y^2$.

give one level of the z-square, leaving a square. This square then folds on one level into the w-square on two levels. Altogether this uses 15 pieces.

Calamity ratchets up with yet a further instance of this instability: There is yet another class, which has the 15-piece folding dissection in Figure 8.12. As in the last dissection, the w-square (pieces K, L, M, N, and O) separates from the x-square. However, piece K will now be on two levels, stealing a square out of the center of piece A. Piece F fills the hole in piece A in the z-square and also is on two levels. To have the top part of piece F to fill in the square hole in piece A with the proper rotational

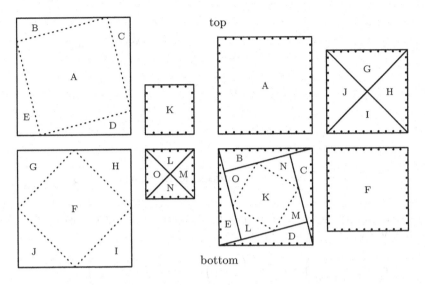

Figure 8.11. Folding squares for $x^2 + y^2 = z^2 + w^2$ where $z = y\sqrt{2}$.

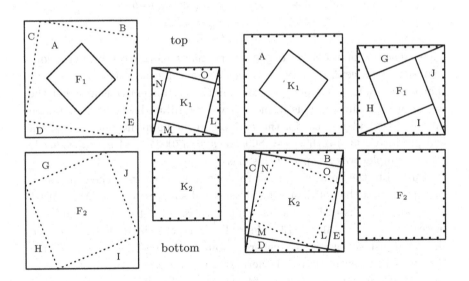

Figure 8.12. Piano-hinged squares for an arctangent relation.

orientation, piece F must rotate in the opposite direction from that of piece A. The three assemblages are on display in Figure 8.13.

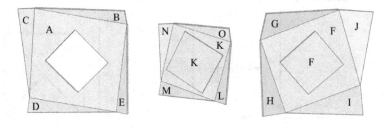

Figure 8.13.   Assemblages of squares for the arctangent relation.

What a strange twist of fortune that makes these angles just so! Let a and b be values, with $0 < a < b$, $a + b < 1$, and

$$\arctan\left(\frac{a}{1-a}\right) + \arctan\left(\frac{b}{1-b}\right) + \arctan\left(\frac{b-a}{1-a-b}\right) = 45°.$$

The dissection will then work for

$$x = \sqrt{a^2 + (1-a)^2}, \quad y = \sqrt{b^2 + (1-b)^2},$$

$$z = 1, \quad w = \sqrt{(b-a)^2 + (1-a-b)^2}.$$

Those readers corrupted by trigonometry may recall with mixed emotions that the *arctangent*, or *inverse tangent*, of a value r is the angle α such that $\tan(\alpha) = r$. The equation involving the arctangents ensures that the top of piece F and the hole in piece A will be rotated so that they match up. The particular values for the example in the figure from $b = 2a$, which will produce $a \approx .142166$, $x \approx .869535$, $y \approx .770082$, and $w \approx .59086$. Is there a simpler way to specify this class?

Pushed to the limit, we retreat to three squares to one. The simplest of such puzzles is that of dissecting three equal squares to one. Abū'l-Wafā found an early dissection that uses nine pieces. Paul Busschop found a 7-piece dissection (Catalan 1873). Henry Perigal (1891) identified a dissection that uses just six pieces. Elsewhere, I found a swing-hingeable dissection that uses just seven pieces. It derives from the use of the T-strip technique. I also found a twist-hingeable dissection that uses just seven pieces, with one of the squares not cut. By applying the conversion technique of Chapter 4, there is then a 13-piece rounded piano-hinged dissection.

This case also succumbs to the T-strip technique, which we have seen in Figures 5.9 and 5.10. That uses 14 pieces but doesn't need to be rounded.

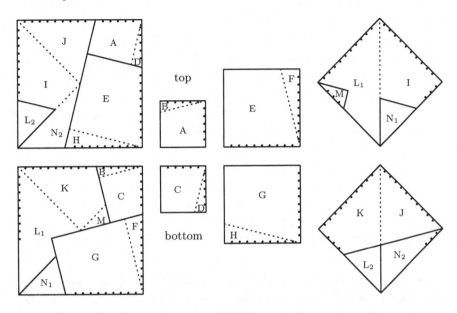

Figure 8.14. Rounded piano-hinged three squares to one, with $x^2 + y^2 = z^2$.

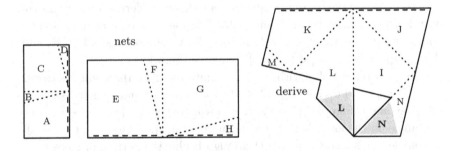

Figure 8.15. Nets and derivation for three squares to one with $x^2 + y^2 = z^2$.

Puzzle 8.3. *Find a 14-piece folding dissection of three identical squares to one that uses the T-strip approach.*

Certain cases of three squares to one are all too susceptible to folding. In particular, there are seismic faults in squares that satisfy the special relationship $x^2 + y^2 = z^2$. I had identified a 7-piece swing-hinged dissection for this case, which suggests the 14-piece rounded piano-hinged dissection in Figure 8.14. We can first create a folding dissection that takes as the top

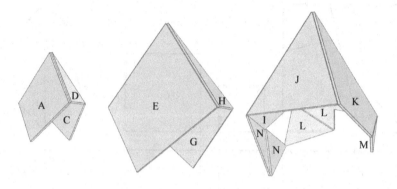

Figure 8.16. Assemblages for three squares to one with $x^2 + y^2 = z^2$.

level of the w-square precisely the 7-piece dissection. We need to reverse the orientations of the x-square and the y-square when we fold them onto the bottom level. When we clip piece M off of piece K and fold it around to fill the notch in piece L, we create a 15-piece folding dissection. The absorption technique produces pieces L and N with portions on both levels.

The nets for the x- and y-squares, as well as the derivation for absorption in the z-square, are in Figure 8.15. The pieces in each of the x-square and the y-square are cap-cyclicly hinged. Furthermore, pieces I, J, K, and L are cap-cyclicly hinged too, as I divulge in Figure 8.16.

Another relationship with tectonic ramifications for three squares to one is $w+x = y+z$. Two R-flaps will produce a 15-piece piano-hinged dissection like the one in Figure 8.17. One R-flap will convert an $(x \times (w - x - y))$-rectangle on the lower left of the w-square to an $((x - (y - x)) \times (z - x))$-rectangle. This involves pieces D through I, plus the portion of piece C on the left. The other converts a $((w - y) \times y)$-rectangle on the upper right of the w-square to a $(z \times x)$-rectangle. This involves pieces J through O, plus the portion of piece C that sticks up. Piece C is the piece that connects these two groups of pieces together, as we can see in Figure 8.18.

As life slowly returns to normal, we realize that a rare, isolated event can unfold quite rapidly. In my first book (1997), I gave a 7-piece unhingeable dissection for squares for $2^2 + (\sqrt{8})^2 + (\sqrt{13})^2 = 5^2$. Figure 8.19 displays a 13-piece folding dissection for the same set of squares. The dissection is primarily an exercise in folding the 2-, $\sqrt{8}$-, and $\sqrt{13}$-squares out flat. The $\sqrt{13}$-square fills the top level, plus one corner of the bottom level, and the remaining two squares fill out the rest of the bottom level. Curiously, we

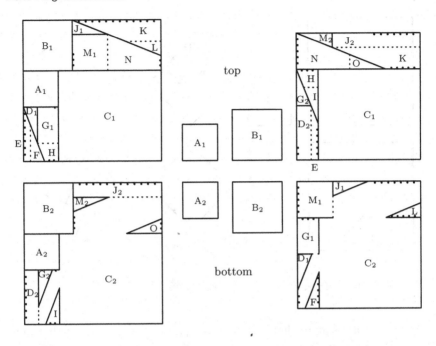

Figure 8.17. Rounded piano-hinged three squares to one with $w + x = y + z$.

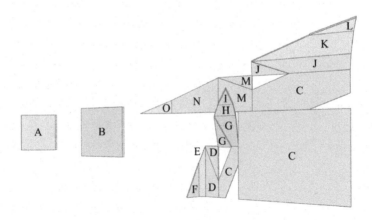

Figure 8.18. Assemblages for three squares to one with $w + x = y + z$.

can just cut the top level of the $\sqrt{8}$-square in the same way as a $\sqrt{8}$-square that Sam Loyd (*Eagle*, 1896a) cut when he gave a 3-piece dissection into a

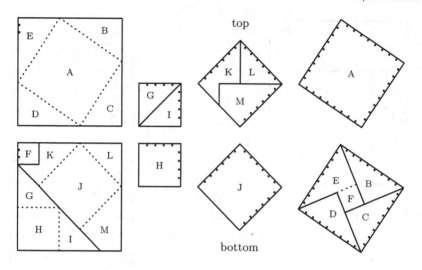

Figure 8.19. Folding dissection of squares for $2^2 + (\sqrt{8})^2 + (\sqrt{13})^2 = 5^2$.

square from a 3-square that has lost a 1-square. Just remember that any time, anywhere, squares may fold on a moment's notice—so, stay vigilant!

Chapter 9

Triangle, Triangle Again

After the last chapter, in which we discovered many techniques that we could apply to general classes of squares, the world of triangles seems a lot tougher. I have found only two general techniques for triangles that beat the conversion methods of Chapters 3 and 4. Yet let's not get discouraged. If at first we don't succeed ..., then let's try triangle after triangle, using trial and error until we triumph. We'll find some special cases worth getting truly excited about and finish the chapter with hexagons, which, since they are actually six triangles glued together, are not to be trifled with.

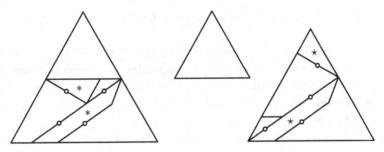

Figure 9.1. Twist-hinged dissection of triangles for $x^2 + y^2 = z^2$.

There are general 5-piece swing-hinged dissections of two unequal triangles to one, which we can convert to 8-piece twist-hinged dissections, which in turn we can convert to 15-piece piano-hinged dissections. However, we can adapt an unhinged dissection by Harry Bradley (1930) into a 6-piece twist-hinged dissection of two equilateral triangles to one (Figure 9.1). Bradley's unhinged dissection works for $x^2 + y^2 = z^2$ whenever $x \leq y$, and so does the twist-hinged adaptation.

We can always convert such a twist-hinged dissection into a 12-piece piano-hinged dissection. Consequently, we ask if there are special relationships for two unequal triangles to one that use fewer pieces. For integer solutions of $x^2 + y^2 = z^2$ where $z = y + 1$ or $z = x + 2$, there are 4-piece swing-hinged dissections, which lead to 6-piece twist-hinged dissections, which in turn lead to 11-piece piano-hinged dissections. Are there any special relationships for which we can find piano-hinged dissections with fewer pieces than those that use our conversion techniques?

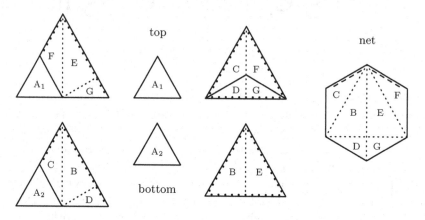

Figure 9.2. Piano-hinged dissection of triangles for $1^2 + (\sqrt{3})^2 = 2^2$.

Taking advantage of the special relationship of $1^2 + (\sqrt{3})^2 = 2^2$, I produce a 7-piece piano-hinged dissection of equilateral triangles in Figure 9.2. There are two cycles of hinges in this dissection. Pieces B, C, E, and F are cap-cyclicly hinged, and pieces B, E, G, and D are flat-cyclicly hinged, as we see in Figure 9.3.

Another special relationship is $(\sqrt{7})^2 + 3^2 = 4^2$, for which I produce an 11-piece piano-hinged dissection of equilateral triangles in Figure 9.4. We see both levels of the 4-triangle laid out in the superposition on the left in Figure 9.5, with most of the boundary between them identified by dotted lines. From the top and bottom vertices, we can identify the area for the 3-triangle, borrowing a flap from the top, which we can then attach to the bottom. This leaves a zigzag band into which we can accommodate a $\sqrt{7}$-triangle (dashed edges), when suitably folded.

The net for the resulting pieces in the $\sqrt{7}$-triangle is on the right in Figure 9.5. There are three cycles of hinges in this dissection. Pieces A, B, C, and D are cap-cyclicly hinged, and pieces D, E, F, and I are flat-

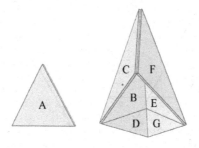

Figure 9.3. Triangles for $1^2 + (\sqrt{3})^2 = 2^2$.

cyclicly hinged, as are pieces F, G, H, and I. We see the two assemblages in Figure 9.6, with the one for the $\sqrt{7}$-triangle shown from its rear.

Pitting two triangles against one may not seem so fair, so let's consider dissecting two triangles into two other triangles. There is a 6-piece

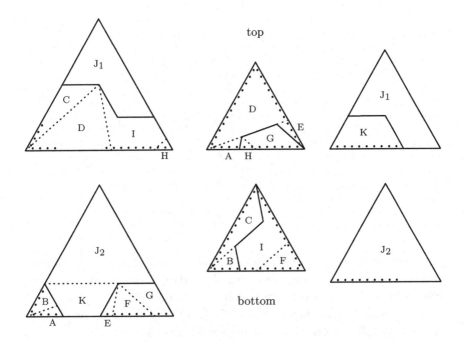

Figure 9.4. Piano-hinged dissection of triangles for $(\sqrt{7})^2 + 3^2 = 4^2$.

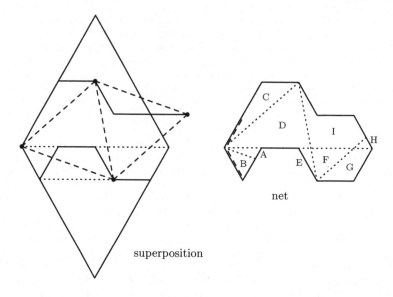

superposition

net

Figure 9.5. Superposition and net for triangles for $(\sqrt{7})^2 + 3^2 = 4^2$.

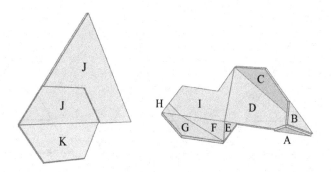

Figure 9.6. Triangles for $(\sqrt{7})^2 + 3^2 = 4^2$.

swing-hinged dissection, which we can convert into a 9-piece twist-hinged dissection and then into a 16-piece piano-hinged dissection. Thus, let's look here for special cases that require fewer than sixteen pieces.

I have found two infinite classes of sets of triangles that give better piano-hinged dissections. The first derives from the identity

$$(2n)^2 + (n+1)^2 = n^2 + (\sqrt{4n^2 + 2n + 1})^2.$$

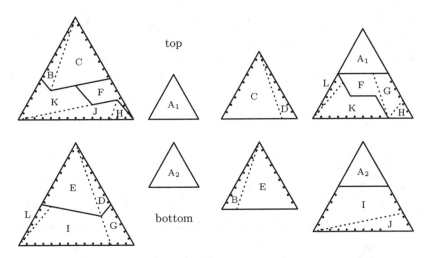

Figure 9.7. Piano-hinged dissection of triangles for $4^2 + 3^2 = 2^2 + (\sqrt{21})^2$. (C)

The last term in this identity represents the square of the distance between certain points in a triangular lattice. For each natural number n, there is a 12-piece piano-hinged dissection. The example in Figure 9.7 is for the case when $n = 2$, or $4^2 + 3^2 = 2^2 + (\sqrt{21})^2$.

We first cut the $(n+1)$-triangle into pieces B, C, D, and E, which will form the apex of the $\sqrt{4n^2 + 2n + 1}$-triangle. We then cut the n-triangle from the apex of the $2n$-triangle. The key is to cut one level of the remaining trapezoidal base of the $2n$-triangle into two groups of pieces, (F, G, H) and (K, L), separated by an oblique step-shaped boundary. The number of

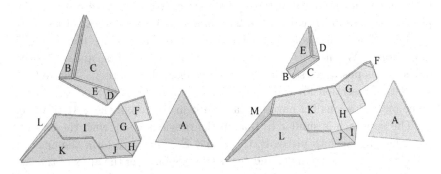

Figure 9.8. $4^2 + 3^2 = 2^2 + (\sqrt{21})^2$.   Figure 9.9. $6^2 + 2^2 = 3^2 + (\sqrt{31})^2$.

(upside-down) steps for pieces F, G, and H is n. As we fold the pieces from the base of the $2n$-triangle to fill in along the base of the $\sqrt{4n^2 + 2n + 1}$-triangle, we see that pieces F, G, and H shift over and down, relative to pieces K and L. The dissection has two cap-cycles and one flat-cycle, as we see in Figure 9.8.

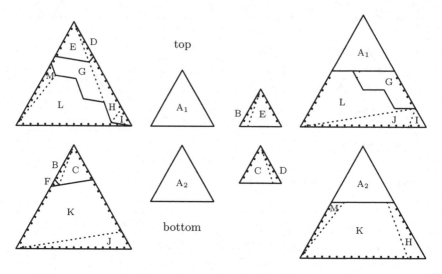

Figure 9.10. Piano-hinged dissection of triangles for $6^2 + 2^2 = 3^2 + (\sqrt{31})^2$.

The second class of sets of triangles derives from the identity

$$(2n)^2 + (n-1)^2 = n^2 + (\sqrt{4n^2 - 2n + 1})^2.$$

Again, the last term in the identity represents the square of the distance between certain points in a triangular lattice. We then have $n - 1$ in the place of $n + 1$, which causes the $2n$-triangle to be the largest of the four triangles. For each whole number $n > 1$, there is a 13-piece piano-hinged dissection. (When $n = 1$, one of the triangles vanishes, and a 7-piece dissection is possible—see the first dissection in this chapter.) Figure 9.10 shows the dissection for the case when $n = 3$, or $6^2 + 2^2 = 3^2 + (\sqrt{31})^2$. Since piece F is small, I have not labeled it in the top of the $\sqrt{31}$-triangle.

The approach is similar to that in the previous class, with the $(n-1)$-triangle splitting into pieces B, C, D, and E, which will form the apex of the $\sqrt{4n^2 - 2n + 1}$-triangle, and the n-triangle forming the apex of the $2n$-triangle. As before, the key is to cut one level of the remaining trapezoidal base of the $2n$-triangle into two groups of pieces, (F, G) and (I, J, L), separated by an oblique step-shaped boundary. The number of steps for pieces I, J, and L is n. As we fold the pieces from the base of the $2n$-triangle to fill in along the base of the $\sqrt{4n^2 - 2n + 1}$-triangle, we see pieces F and G shift over and up, relative to pieces I, J, and L. This different direction for the shift results in one more piece. Again, the dissection has two cap-cycles and one flat-cycle, as we see in Figure 9.9.

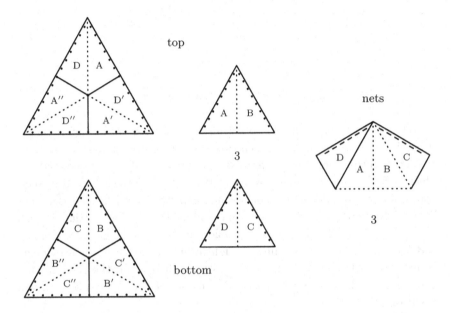

Figure 9.11.   Folding dissection of three equal triangles to one.

It's time to handle our three-sided figures in threesomes, i.e., three triangles to one. Proudly, we submit our first one in triplicate: three congruent triangles identically cut and hinged. For the unhinged version, Plato described a 6-piece dissection in *Timaeus*. We can produce a folding variation of this dissection by using the unhinged version on each of the two levels and then hinging the pieces appropriately. This gives us the 12-piece folding dissection in Figure 9.11. A perspective view of the three identical assemblages is in Figure 9.12, close to the orientation that you would see as you flatten the assemblages to form the large triangle.

As we can see in the perspective view, each small triangle is cap-cyclicly hinged. There is another way to hinge the pieces, which is flat-cyclic.

Puzzle 9.1. *Find the flat-cyclic hinging of the pieces in Figure 9.11.*

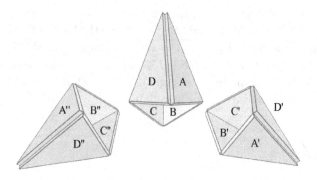

Figure 9.12. Perspective view of three equal triangles to one.

When our trio of triangles are not all congruent, we have a wide range of possibilities. In particular, when two of the triangles are congruent and smaller than the third, we look for dissections that use fewer than 16 pieces, which is what we can achieve by converting a 9-piece twist-hinged dissection, which we would get from converting a 6-piece swing-hinged dissection. Figure 9.13 displays a nifty 9-piece piano-hinged dissection of triangles for $1^2 + 1^2 + (\sqrt{7})^2 = 3^2$. The trick is to fold out the $\sqrt{7}$-triangle into a shape that fills out most of the outline of a folded-out 3-triangle. This is evident from the net of the $\sqrt{7}$-triangle assemblage, which is on the right in the same figure. The assemblage has both a cap-cycle and a flat-cycle, as we see in Figure 9.14.

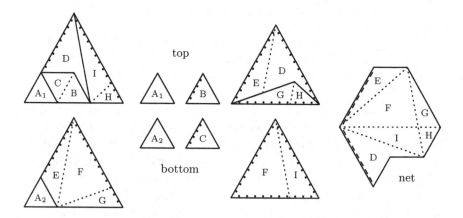

Figure 9.13. Piano-hinged dissection of triangles for $1^2 + 1^2 + (\sqrt{7})^2 = 3^2$.

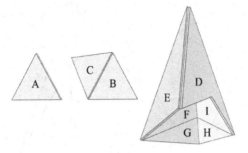

Figure 9.14. Perspective view of triangles for $1^2 + 1^2 + (\sqrt{7})^2 = 3^2$.

Another special case is for $1^2 + (\sqrt{3})^2 + (\sqrt{3})^2 = (\sqrt{7})^2$. I have found the 16-piece folding dissection in Figure 9.15. Primed letters indicate the pieces from the second of the $\sqrt{3}$-triangles in the $\sqrt{7}$-triangle. As with the last several dissections, we can derive this dissection by superposing tessellations (on the right in Figure 9.16). Unfold the 1-triangle into two pieces,

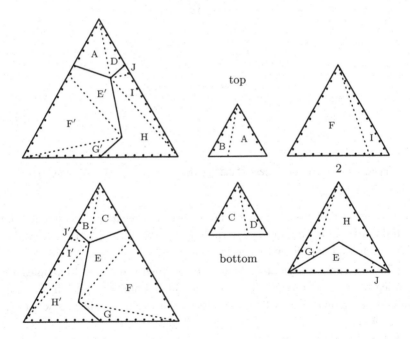

Figure 9.15. Folding dissection of triangles for $1^2 + (\sqrt{3})^2 + (\sqrt{3})^2 = (\sqrt{7})^2$.

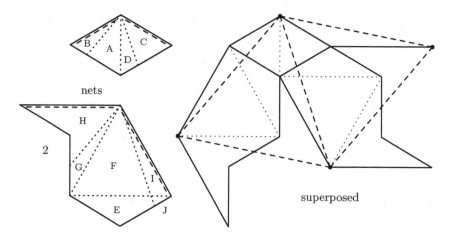

Figure 9.16. Nets and superposition for triangles for $1^2 + (\sqrt{3})^2 + (\sqrt{3})^2 = (\sqrt{7})^2$.

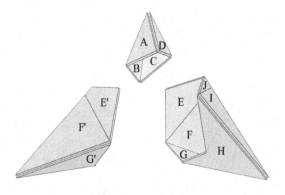

Figure 9.17. Perspective view of triangles for $1^2 + (\sqrt{3})^2 + (\sqrt{3})^2 = (\sqrt{7})^2$.

and unfold each of the $\sqrt{3}$-triangles into three pieces, as I indicate with the dotted lines. Arrange and overlay these pieces with the $\sqrt{7}$-triangle, which I have unfolded and shown with dashed edges.

The superposition yields a dissection in which the two $\sqrt{3}$-triangles are identically cut and hinged. Each of the three small triangles has a cap-cyclic hinging, and each $\sqrt{3}$-triangle has a flat-cyclic hinging, as we see in Figure 9.17.

When no two of the three triangles are congruent, we look for dissections that use fewer than 25 pieces, which is what we can achieve by converting

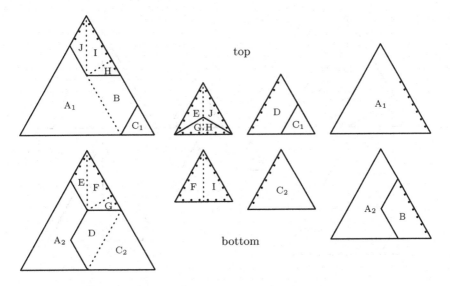

Figure 9.18. Piano-hinged dissection of triangles for $(\sqrt{3})^2 + 2^2 + 3^2 = 4^2$.

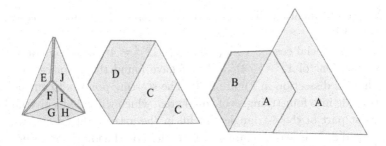

Figure 9.19. Perspective view of triangles for $(\sqrt{3})^2 + 2^2 + 3^2 = 4^2$.

a 13-piece twist-hinged dissection, which we would get by converting an 8-piece swing-hinged dissection. As an example, we have a relatively easy 10-piece piano-hinged dissection of triangles for $(\sqrt{3})^2 + 2^2 + 3^2 = 4^2$ in Figure 9.18. We can position the 3-triangle in the left corner of the 4-triangle and then fill in the opposite side of the 4-triangle with the other two triangles, suitably folded. The $\sqrt{3}$-triangle folds out to form a hexagon and then together for a trapezoid. We could use four pieces to fold the 2-triangle into a parallelogram half of its height, but we do better by folding pieces B and D out of the 3-triangle and 2-triangle, respectively, in a manner

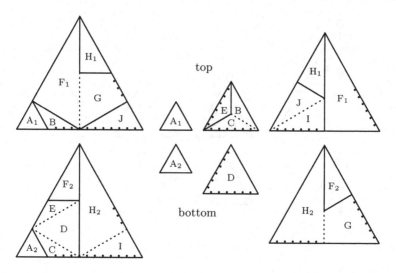

Figure 9.20. Rounded piano-hinged triangles for $1^2 + (\sqrt{3})^2 + (\sqrt{12})^2 = 4^2$. **(C)**

which is complementary. The $\sqrt{3}$-triangle is cap-cyclicly hinged, as we see in Figure 9.19.

A rather special case is $1^2 + (\sqrt{3})^2 + (\sqrt{12})^2 = 4^2$, which is really just two applications of $1^2 + (\sqrt{3})^2 = 2^2$. I have found the 10-piece rounded piano-hinged dissection in Figure 9.20. The starting point is to cut up the $\sqrt{12}$-triangle into four triangles of equal area, which will fold to give all but the lowest part of the 4-triangle. To fill in the lower right corner, we cut pieces I and J. This leaves a larger area to fill in on the lower left corner, for which we use the 1- and $\sqrt{3}$-triangles. Once we have done this, however, we see that we can apply absorption, involving pieces F and H, reducing what would have been an 11-piece dissection to ten pieces. We see a perspective view of the assemblages in Figure 9.21.

Another special case is $1^2 + 2^2 + (\sqrt{7})^2 = (\sqrt{12})^2$. I have found the 18-piece folding dissection in Figure 9.22. Imagine unfolding the 2-triangle into a rhombus that will span between two vertices of the $\sqrt{12}$-triangle. Then, cut the $\sqrt{7}$-triangle so that it will fill out all but the bottom two corners of the $\sqrt{12}$-triangle. To avoid splitting the 1-triangle into two assemblages, we must be a bit more clever in unfolding the 2-triangle, so that it fills one of the two bottom corners and leaves a hole twice as large for the other corner. This is actually a superposition technique. We see the superposition and the nets in Figure 9.23. Each of the three triangles is cap-cyclicly hinged,

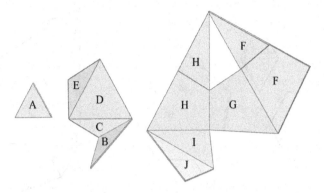

Figure 9.21. Perspective view of triangles for $1^2 + (\sqrt{3})^2 + (\sqrt{12})^2 = 4^2$.

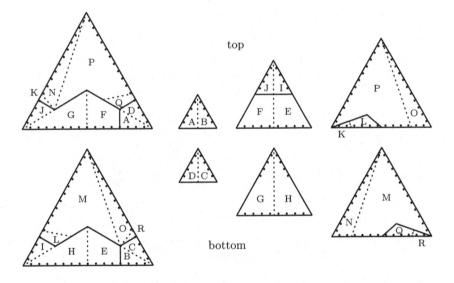

Figure 9.22. Folding dissection of triangles for $1^2 + 2^2 + (\sqrt{7})^2 = (\sqrt{12})^2$.

and there are two more flat-cycles in the $\sqrt{7}$-triangle. I have arranged the three assemblages in Figure 9.24 to suggest how they would fit together to form the $\sqrt{12}$-triangle.

We have seen a number of special cases of triangles for $x^2 + y^2 + z^2 = w^2$ in which two of the side lengths are whole numbers. Consequently, the reader can hardly complain about solving a puzzle of the same type:

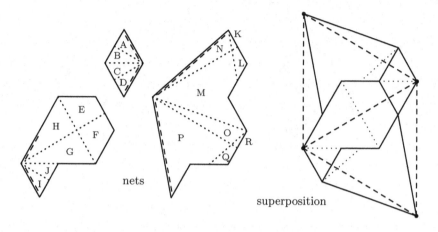

Figure 9.23. Nets and superposition for triangles for $1^2 + 2^2 + (\sqrt{7})^2 = (\sqrt{12})^2$.

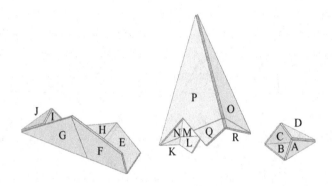

Figure 9.24. Perspective view of triangles for $1^2 + 2^2 + (\sqrt{7})^2 = (\sqrt{12})^2$.

Puzzle 9.2. *Find a folding dissection of triangles for $1^2 + (\sqrt{3})^2 + 3^2 = (\sqrt{13})^2$ that uses at most 18 pieces.*

We next attack triangles for $1^2 + (\sqrt{7})^2 + (\sqrt{13})^2 = (\sqrt{21})^2$, for which we can find the 19-piece solution in Figure 9.25. Once again, the superposition technique comes to our rescue, as we see in Figure 9.26. We fold out the 1-triangle into two pieces, the $\sqrt{13}$-triangle into four pieces, and the $\sqrt{7}$-triangle into five pieces, as we indicate with the dotted lines. Dashed edges indicate the boundaries of the $\sqrt{21}$-triangle.

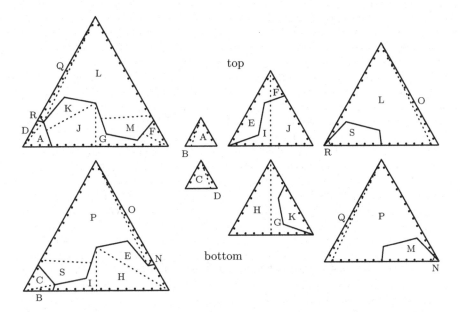

Figure 9.25. Folding dissection of triangles for $1^2 + (\sqrt{7})^2 + (\sqrt{13})^2 = (\sqrt{21})^2$.

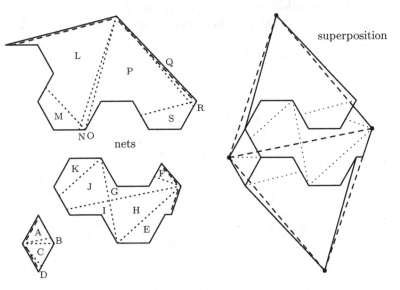

Figure 9.26. Nets and superposition for $1^2 + (\sqrt{7})^2 + (\sqrt{13})^2 = (\sqrt{21})^2$.

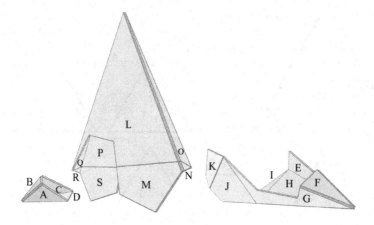

Figure 9.27. Assemblages of triangles for $1^2 + (\sqrt{7})^2 + (\sqrt{13})^2 = (\sqrt{21})^2$.

We need real teamwork between the triangles to pull off this triple play. Folding out two of the pieces of the $\sqrt{13}$-triangle makes space for the pieces of the $\sqrt{7}$-triangle, which we also seek to economize. Since $\sqrt{21} = \sqrt{3} \times \sqrt{7}$, a 2-piece folding out of the $\sqrt{7}$-triangle would span between two vertices of

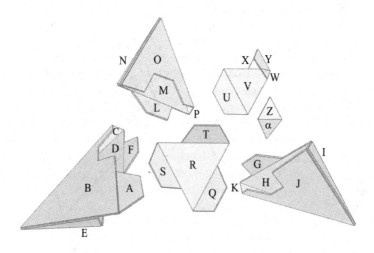

Figure 9.28. Assemblages: triangles for $1^2 + 2^2 + 3^2 + 4^2 + 5^2 + 6^2 = (\sqrt{91})^2$.

the $\sqrt{21}$-triangle. To accommodate both the 1-triangle and the $\sqrt{7}$-triangle, I transfer the area of one face of the 1-triangle from the left end of a 2-piece folding out to the right end. This then allows each of the small triangles to contribute all of the pieces incident on one vertex in the $\sqrt{21}$-triangle, meaning that each of the three small triangles has a cap-cycle. Also the $\sqrt{7}$- and $\sqrt{13}$-triangles each has a flat-cycle. Again, I have arranged the three assemblages in Figure 9.27 to suggest how they would fit together to form the $\sqrt{21}$-triangle.

Ernest Freese (1957b) noticed that the sum of the squares of the first six natural numbers is 91 and that the $\sqrt{91}$ is the distance between two of the points of a triangular lattice. This enabled him to find an 11-piece unhingeable dissection of the corresponding set of triangles. How about a

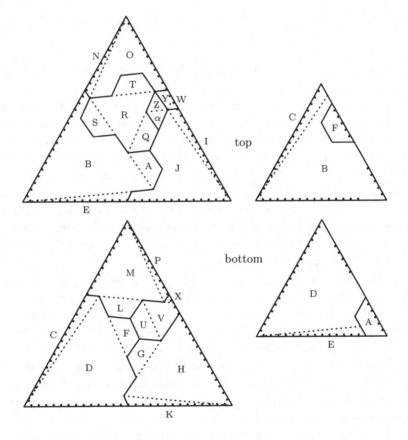

Figure 9.29. Folding of triangles for $1^2 + 2^2 + 3^2 + 4^2 + 5^2 + 6^2 = (\sqrt{91})^2$ started.

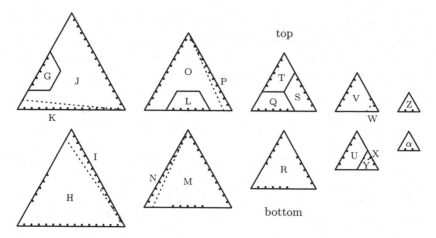

Figure 9.30. Folding triangles for $1^2 + 2^2 + 3^2 + 4^2 + 5^2 + 6^2 = (\sqrt{91})^2$ completed.

folding dissection for these triangles? That is a bit more difficult, but I found the 27-piece dissection in Figures 9.28, 9.29, and 9.30.

The basic approach is to fill each of the three corners of the $\sqrt{91}$-triangle with one of the 4-, 5-, and 6-triangles. This creates an overlap between the 5- and 6-triangles, which we resolve by cutting out pieces A and G. Fitting the 1-, 2-, and 3-triangles in the remaining space is a bit tricky. The key seems to be to cut the 3-triangle so that we can fold out pieces Q, S, and T and cut pieces F and L out of the 6- and 4-triangles, respectively, to accommodate the folded-out 3-triangle. This leaves an area into which the 1- and 2- triangles can be folded. I have arranged the six assemblages in Figure 9.28 to suggest how they would fit together to form the $\sqrt{91}$-triangle. What a three-ring circus!

Harry Lindgren (1964b) considered the case of dissecting hexagons for $1^2 + (\sqrt{3})^2 = 2^2$. He gave a 6-piece dissection that is swing-hingeable. I have discovered how to adapt his basic dissection to give the 13-piece folding dissection in Figure 9.32. Use his dissection on the bottom level, and on the top level of the $\sqrt{3}$-hexagon, use his dissection but cut the large piece into two pieces that we can fold around. A nifty feature of this dissection is that there are three cap-cyclic hingings, which makes for the lovely three-dimensional view in Figure 9.31. Breaking these cyclic hingings, we get the nets in Figure 9.33.

Just as the dissection of triangles in Figure 9.2 is a special case relative to the second class of dissections of two triangles to two triangles, so too

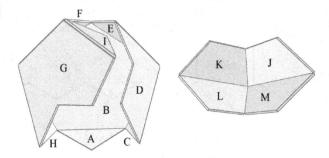

Figure 9.31.   View of the assemblages for hexagons for $1^2 + (\sqrt{3})^2 = 2^2$.

is the preceding dissection of hexagons a special case relative to a corresponding class of dissections of two hexagons to two hexagons. Therefore, based on the identity

$$(2n)^2 + (n-1)^2 = n^2 + (\sqrt{4n^2 - 2n + 1})^2,$$

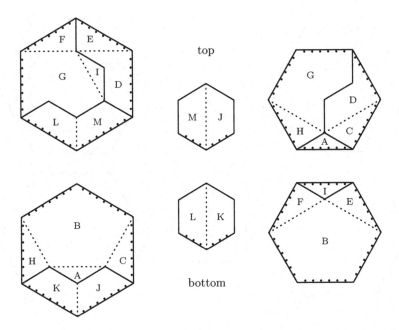

Figure 9.32.   Folding dissection of hexagons for $1^2 + (\sqrt{3})^2 = 2^2$. **(C)**

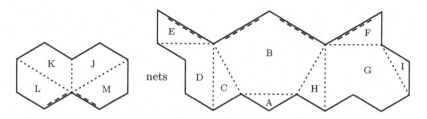

Figure 9.33. Nets for hexagons for $1^2 + (\sqrt{3})^2 = 2^2$.

for each whole number $n > 1$, there is a 22-piece piano-hinged dissection. Figure 9.34 shows the dissection for the case when $n = 3$, or $6^2 + 2^2 = 3^2 + (\sqrt{31})^2$.

The approach is somewhat related to the approach for triangles, in that in going from the $\sqrt{4n^2 - 2n + 1}$-hexagon to the $2n$-hexagon, we shift pieces over by using the step approach. However, instead of perching the $(n-1)$-triangle atop the $\sqrt{4n^2 - 2n + 1}$-triangle, and the n-triangle atop the $2n$-triangle, we can carve the $(n-1)$-hexagon out of the n-hexagon

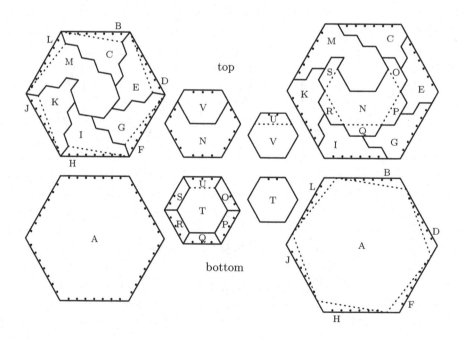

Figure 9.34. Folding dissection of hexagons for $6^2 + 2^2 = 3^2 + (\sqrt{31})^2$.

and hide the remainder of the n-hexagon in the center of the $2n$-hexagon. I have glued one level of an $(n-1)$-hexagon onto piece M, so that the $\sqrt{4n^2 - 2n + 1}$-hexagon needs no part of either the $(n-1)$-hexagon or the n-hexagon. We are left with a lovely, if only partial, hexagonal symmetry. However, as we can see in Figure 9.35, cyclic hingings are now nowhere to be found.

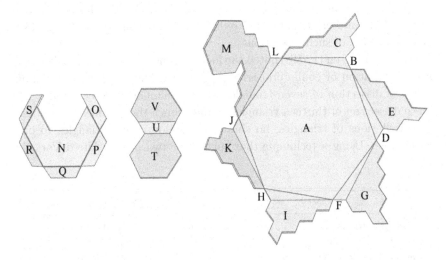

Figure 9.35. View of the assemblages for hexagons for $6^2 + 2^2 = 3^2 + (\sqrt{31})^2$.

Having found our third infinite class of dissections is such a perfect way to conclude our chapter—like winning the Triple Crown! It has been a long, tough battle that has finally yielded to a three-pronged assault: the special case at the beginning of the chapter, the two classes of triangles in the middle of the chapter, and the class of hexagons at the end of the chapter. So set up your tripods, and snap all the pictures you want of our triple winners.

Manuscript 3

Giving Freese His Due

It was a revelation to learn that Ernest Freese had described many dissections in his manuscript that others rediscovered after his death. Table M3.1 is a partial list of such dissections.

Harry Lindgren (1964b) noted the fact that the $\sqrt{7}$ appears as a distance in a tessellation of equilateral triangles, so that there is a corresponding 13-piece dissection of seven triangles to one. Similarly, we can find a 22-piece dissection of thirteen triangles to one. Since the $\sqrt{5}$ does not appear in tessellations of triangles, no similar dissection of five triangles to one is possible. Using a technique that turns over two pieces, Alfred Varsady

Dissection	Pieces	Competitor
five $\{3\}$s to one	9	Varsady (1989)
seven $\{3\}$s to one	12	Collison (1980s)
seven $\{3\}$s to three	21	Paterson (1989)
$\{4\}$s for $9^2 + 12^2 + 20^2 = 25^2$	5	Frederickson (1997)
two $\{4\}$s to two other $\{4\}$s	6	Frederickson (1997)
five $\{4\}$s to two $\{4\}$s	12	Paterson (1995)
four $\{5\}$s to one	6	Lindgren (1964)
one $\{6\}$ to one $\{5\}$	7	Lindgren (1964)
two $\{6\}$s to one $\{3\}$	6	Lindgren (1964)
two $\{6\}$s to one	9	Lindgren (1964)
three $\{6\}$s to one	6	Lindgren (1964)
seven $\{6\}$s to one	12	Lindgren (1964)
three $\{6/2\}$s to one	12	Lindgren (1964)
eight $\{8\}$s to one	24	Lindgren (1964)
one $\{10\}$ to two $\{5\}$s and two $\{5/2\}$s	6	Varsady (1980s)
one $\{10/3\}$ to two $\{5/2\}$s *	10	Lindgren (1964)
one $\{12\}$ to one $\{3\}$	8	Lindgren (1964)
one $\{12\}$ to three $\{4\}$s	9	Elliott (1986)
one $\{G\}$ to one $\{3\}$	5	Lindgren (1961)
one $\{G(\alpha)\}$ to one $\{4\}$	5	Frederickson (2002)

* not included in *Geometric Transformations*.

Table M3.1. Ernest Freese's discoveries (1952–1957) before others.

(1989) had found a 9-piece dissection. Yet, thirty years earlier, Freese had found a 9-piece dissection (Figure M3.1) with no pieces turned over.

Figure M3.1. Freese's five triangles to one.

Freese relied on the P-slide technique. As shown on the left in Figure M3.2, he formed a parallelogram from four of the triangles. His goal was to create a trapezoidal base on which to set the fifth triangle. He then applied the P-slide (dotted edges) to produce a parallelogram with the same height as the trapezoidal base. Once he had a parallelogram of the desired height (on the right in Figure M3.2), he cut a triangle off the right-hand side and rotated it down (dotted edges) to give the trapezoid.

In hindsight, Freese's dissection looks simple. However, the key was to convert the parallelogram to the trapezoid *after* performing the P-slide. Reversing the order of the operations creates more pieces. Freese used the same basic idea to find a 12-piece dissection of seven triangles to one and a 20-piece dissection of thirteen triangles to one. Both of these dissections improve on the tessellation approach alluded to above.

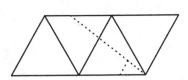 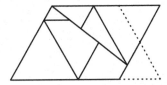

Figure M3.2. Derivation of Freese's five triangles to one.

The 9-piece dissection of two hexagons to one (Figure M3.3) is another example. Based on tessellations, the approach also works for unequal hexagons. For a final example, enjoy the dissection of three hexagrams to one in Figure M1.2 (in my first section on Freese's manuscript).

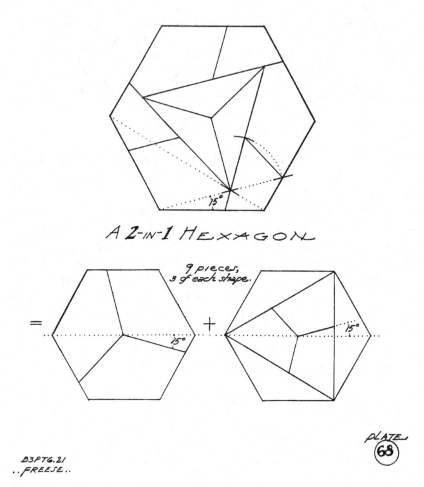

A 2-IN-1 HEXAGON

9 pieces,
3 of each shape.

B3PT6.21
..FREESE..

PLATE
68

Figure M3.3.   Ernest Freese's Plate 68.

Chapter 10

Hexagons, Triangles, Squares—Oh My!

In this chapter we encounter the fearsome Big Three in the land of "Onz": hexagons, triangles, and squares. And they will appear in quantities that will make us gasp—*oh, my!* We shall first look to the tenth-century Arabic-Islamic mathematician Abū'l-Wafā, surely a geometric wizard, to show us the road to follow. He described a magical, though unhinged, dissection of $a^2 + b^2$ squares to one, where a and b are natural numbers. Following his lead, the twentieth-century Australian patent examiner Harry Lindgren and the Los Angeles architect Ernest Freese joined in, performing a similar wizardry on $a^2 + ab + b^2$ triangles and on $a^2 + ab + b^2$ hexagons. Adapting these techniques, and bringing in others too, we will see how to make these polygonal beasts bow on our command.

Let's summon up our courage and begin with squares. Of course, when $a = 1$ and $b = 1$, we have the dissection of two squares to one (Figure 2.6). With five squares to one, we find a more interesting application of Abū'l-Wafā's method. We can easily adapt his dissection to a folding dissection. The bottom level of the large square is just the mirror image of the top level, and we fold-hinge appropriately to get the 18-piece dissection in Figure 10.1. As we see in the perspective view of Figure 10.2, the three identical small squares are flat-cyclicly hinged. A different arrangement and hinging of the pieces will replace the flat-cycles with cap-cycles.

A similar approach works with ten squares to one, with $a = 1$ and $b = 3$. The 32-piece dissection in Figure 10.3 has the four identical small squares with pieces G, H, I, and J flat-cyclicly hinged. The two small squares with pieces C, D, E, and F are cap-cyclicly hinged, as we see in Figure 10.4. Again, other arrangements and hingings of the pieces are possible.

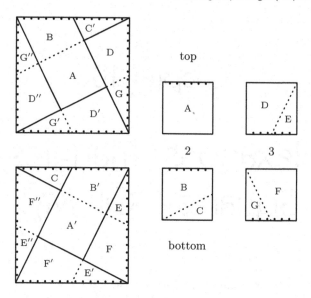

Figure 10.1. Folding dissection of five squares to one.

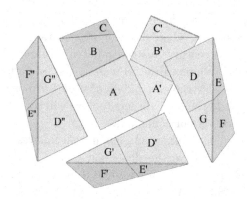

Figure 10.2. Perspective of folding five squares to one.

So far, our dissections of equal squares have been rather tame, more like teddy bears than grizzlies. However, when we get to five squares to two, we begin to hear them growl. If we were willing to settle for an unhingeable dissection of five squares to two, then tessellations would do just fine. Ernest Freese (1957b), David Paterson (1995), and Robert Reid all found 12-piece dissections. However, a straightforward application of the

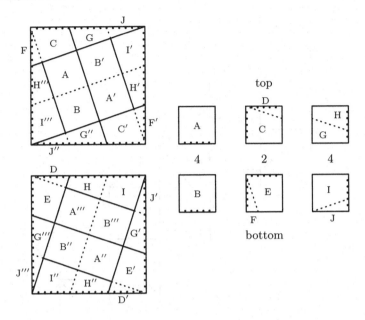

Figure 10.3.   Folding dissection of ten squares to one.

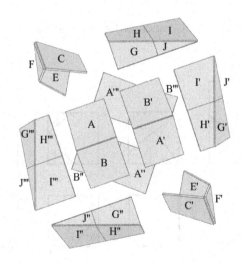

Figure 10.4.   Perspective of folding ten squares to one.

tessellation technique seems to fail us when we look for a folding dissection.
Instead, let's try the strip technique in Figure 10.5. We crosspose a strip

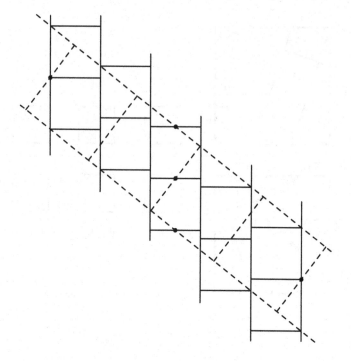

Figure 10.5.   Crossposition for five squares to two.

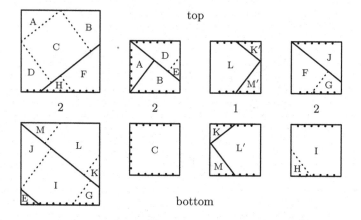

Figure 10.6.   Folding dissection of five squares to two.

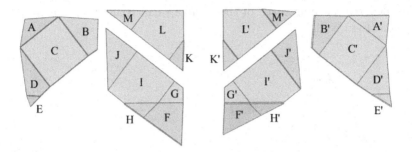

Figure 10.7. Perspective of folding five squares to two.

whose element consists of two large squares with a strip whose element consists of five small squares. The crossposition is similar to that for three squares to one in Solution 8.3 (see Chapter 20).

The resulting 26-piece dissection lumbers forth in Figure 10.6. We need the T-strip method rather than the P-strip for two reasons. First, we need to cut the middle of the five small squares symmetrically, and our T-strip does this. Second, we need to use the alternate fold-hinging of the pieces at the ends of the strip of large squares. Thus, we attach piece A to piece C; there is no other choice. Pieces F, G, H, and I are flat-cyclicly hinged, as we see in Figure 10.7.

Eventually, I did find an attractive folding dissection of five squares to two that starts with tessellations.

Puzzle 10.1. *Find a 26-piece folding dissection of five squares to two that derives from tessellations.*

Having brought squares under control, we turn our attention to triangles. Guided by $a^2 + ab + b^2$, we observe that three triangles to one (Figure 9.11) covers the case when $a = 1$ and $b = 1$. For $a = 1$ and $b = 2$, Lindgren (1964b) used tessellations to find a 13-piece dissection of seven triangles to one. We can readily adapt it to a 26-piece folding dissection (Figure 10.8). We need to cut two of the small triangles so that one level of each fills in the center of one level of the large triangle. Then, we cut two small triangles identically so that we can fill in the middle portions of the remaining two sides of the large triangle. We can hinge each of these two small triangles in a flat-cyclic fashion. Finally, we cut the remaining three small triangles identically so that we can fill in the corners of the large triangle. We can hinge each of these three small triangles in a cap-cyclic fashion (Figure 10.9).

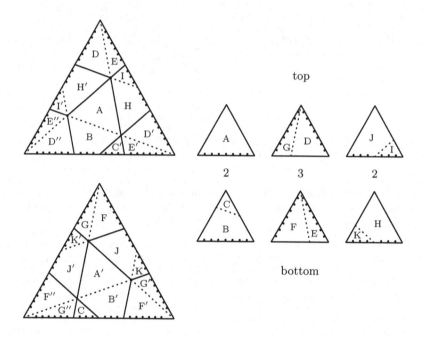

Figure 10.8. Folding dissection of seven triangles to one.

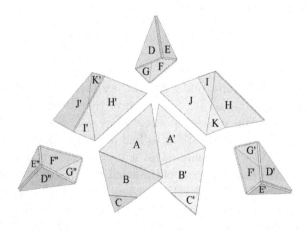

Figure 10.9. Perspective view of seven triangles to one.

Again, a reader might complain that this dissection of triangles is more like a pussycat than a jungle cat. Less likely to produce a faint meow is the wild seven triangles to three triangles. I have found the 45-piece piano-hinged dissection in Figure 10.10. That may seem like a lot of pieces, but 21 is the fewest pieces known for an unhingeable dissection. And that dissection, described by Freese (1957b) and Paterson (1989), gives no hint of how to create a piano-hinged dissection.

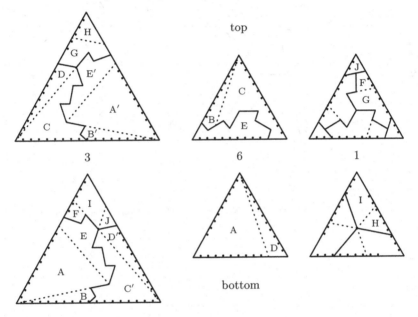

Figure 10.10. Rounded piano-hinged dissection of seven triangles to three. (C)

A superposition (on the left in Figure 10.11) leads us to the dissection. Fold out each of the seven small triangles into a flat figure: one into a hexagon, and the other six into what would have been hexagons if we had not sliced a piece off one side and glued it onto another. Then, arrange these six figures around the hexagon. Next, unfold the three large triangles in the simplest fashion, arrange them to form a hexagon, and superpose the figure created from the small triangles.

However, we are not yet done. Each of the six identical folded-out triangles has a small irregular triangle, which we can eliminate if we add and subtract copies of this small triangle from various pieces. We see on the right of Figure 10.11, with the copies of triangles added and subtracted from the pieces, what was the central hexagon and one of the six identical folded-

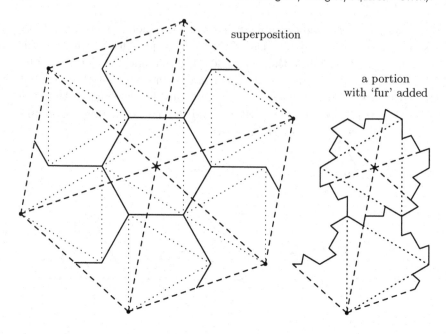

superposition

a portion
with 'fur' added

Figure 10.11. Preliminary superposition and 'fur' for seven triangles to three.

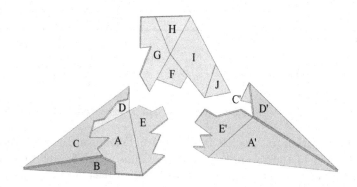

Figure 10.12. Some of the assemblages for seven triangles to three.

out triangles. Do not be frightened by the remarkable "furry" appearance produced by this addition and subtraction process. Rather, admire its fearful symmetry. A perspective view of the assemblages for one of the large triangles appears in Figure 10.12. Each of the six identical assemblages has cap-cyclic hinging, and each of the three assemblages from the seventh small triangle has a flat-cyclic hinging. Although different from Lindgren's approach, this is still a technique based on tessellations.

If we proceed to a dissection of thirteen triangles to one ($a = 1$ and $b = 3$), we must be quick on our feet to avoid getting mauled. Our adaptation of Lindgren's approach will give a 44-piece folding dissection, in contrast to the 43 pieces that we would get by converting a 21-piece swing-hinged dissection to a 26-piece twist-hinged dissection, and then from that to a piano-hinged dissection.

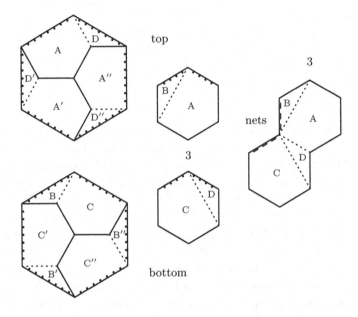

Figure 10.13. Folding dissection of three hexagons to one. (C)

Hexagons share many similarities with triangles. Lindgren (1964b) showed how to use tessellations to find dissections of $a^2 + ab + b^2$ hexagons to one. Freese (1957b) anticipated him in the cases of three hexagons to one ($a = b = 1$) and seven hexagons to one ($a = 2, b = 1$). We can easily adapt their 6-piece dissection of three hexagons to one to give a 12-piece folding dissection (Figure 10.13).

We cut and hinge each of the small hexagons identically, producing a large hexagon with rotational symmetry. We can hinge each small hexagon in a cap-cyclic fashion, as we see in Figure 10.14.

Lindgren (1964b) and Freese (1957b) both found the same 12-piece dissection of seven hexagons to one. Unfortunately, it does not seem to

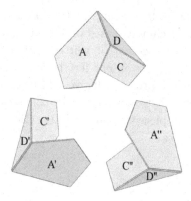

Figure 10.14. Perspective view of hexagons for three hexagons to one.

adapt readily to a folding dissection. However, a 13-piece variation that I
identified does adapt to give a 26-piece folding dissection (Figure 10.15).
This is the same number of pieces as for seven triangles to one. We need to
cut two of the small hexagons, so that one level of each fills in the center
of the two levels of the large hexagon. We can then cut the remaining five
small hexagons identically, so that we can fill in the remaining five corners
of the large hexagon. We can hinge each of these five small hexagons in a
cap-cyclic fashion, as we see in Figure 10.16.

There is a second way to fold-hinge the underlying dissection in Figure 10.15, assuming that we ignore the labels of the pieces and alter the
relative orientation of top and bottom levels in some of the small hexagons.
The pieces will end up flat-cyclicly rather than cap-cyclicly hinged.

Puzzle 10.2. *Find a 26-piece flat-cyclic hinging of seven hexagons to one.*

Before you complain that the last hexagon dissection is just a "dandy-
lion," let's see if you have the nerve to confront a beast that will let out a
real roar. Can you handle a dissection of seven hexagons to three? Freese
(1957b) and Lindgren (1964b) used the tessellation approach to discover
an 18-piece unhingeable dissection. That dissection gives only the most
general of hints as to how to proceed, such as cutting the seventh small
hexagon into three equal parts and having some vertex of each of the other
small hexagons positioned at a vertex of one of the three large hexagons.
If you follow these hints in a straightforward fashion, you find yourself
cornered—unable to fold several of the pieces into their intended positions.

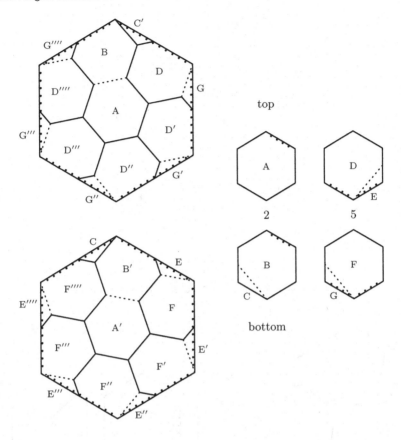

Figure 10.15. Folding dissection of seven hexagons to one.

The 44-piece rounded piano-hinged dissection in Figure 10.17 then begins to look rather good.

First, a warning about interpreting the figure: I have shown the pieces for one of the three large hexagons, with the other two being slight variations. The second has pieces O, P, and Q rather than pieces M and N, and the third has pieces R, S, and T rather than pieces M and N.

How did I spring my trap on this dissection? I started with a large hexagon from Freese's and Lindgren's 18-piece dissection. This corresponds to the top level of a large hexagon in Figure 10.17. I used the mirror image for the bottom level, with pieces A, C, and K corresponding to pieces J, I, and D, respectively. I found a way to fold five of the pieces into one small hexagon and another five into a second small hexagon. However, this left

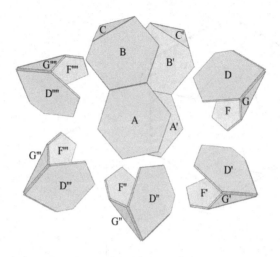

Figure 10.16. Perspective view of seven hexagons to one.

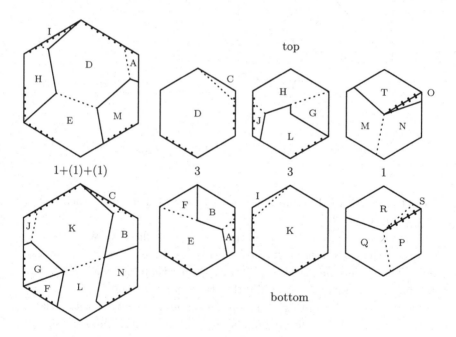

Figure 10.17. Rounded piano-hinged dissection of seven hexagons to three. **(C)**

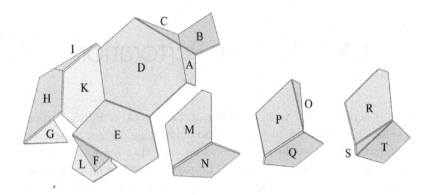

Figure 10.18. Perspective view for seven hexagons to three.

two disconnected holes to be filled by just one assemblage. (These holes correspond to piece M on the top level and the union of pieces F and G on the bottom level.) I filled the hole on the bottom level with pieces F and G, stealing them from originally larger versions of pieces B and L, respectively. The new cavity on the right in the lower level accommodates piece N. All but the last small hexagon has a flat-cyclic hinging. The positions of the three assemblages on the left in Figure 10.18 suggest how one of the large hexagons fits together. If we replace the assemblage comprised of pieces M and N with either of the two assemblages to its right, we get one of the remaining large hexagons.

It wasn't always easy, but we have found the heart, the brains, and the nerve to face down some truly ferocious creatures in this chapter. Let's hold our heads high as we make our escape from this wild menagerie and head for home.

Folderol 2

One Percent Perforation

Almost everyone seems to have heard of Thomas Alva Edison's remark that genius is one percent inspiration and ninety-nine percent perspiration. Most puzzle-solvers would probably agree with that sentiment. On the other hand, few people have heard of postage-stamp folding puzzles, with which we will have some fun here. Of course, any puzzle comes with its own unique blend of enjoyment and challenge, and postage stamp puzzles are no exception. In a nod to Edison, we could characterize them as ninety-nine percent infuriation and one percent perforation.

For the benefit of future generations, so used to sending communications electronically, here are a few particulars about stamps. Before the ubiquitous use of electronic networks, a person would write or type a message on sheets of paper, slip them inside an envelope, write the physical address on the outside, and "post" the envelope. If no postage meter was in use, the person would glue (often decorative) postage stamps on the outside of the envelope to pay the postal service to deliver that envelope. The stamps, almost always rectangular in shape, were printed by the postal service on sheets of paper, producing $n \times m$ blocks of stamps. Before peel-off sheets of stamps were introduced, stamps had a saliva-soluble adhesive on the back and were separated by rows of small round holes, called perforations. A person would tear off a stamp along the perforations on its boundary, lick the adhesive on the back, and glue the stamp onto the envelope.

Let us travel back in time to that heyday of the stamp, the twentieth century. Henry Dudeney featured two stamp-folding puzzles in his puzzle column "Perplexities" (*Strand*, 1920a). Subsequently, he included the puzzle in his book *Modern Puzzles and How to Solve Them* (1926). Here is the original version:

490.–FOLDING POSTAGE-STAMPS.

If you have eight postage-stamps, 4 by 2, as in the diagram, it is very interesting to discover the various ways in which they can be folded so that they all lie under one stamp, as shown. I will say at once that they can actually be folded in forty different ways so that No. 1 is face upwards and all the others invisible beneath it. Nos. 5, 2, 7, and 4 will always be face downwards, but you can arrange for any stamp, except No. 6, to lie next to No. 1, though there are only two ways each in which 7 and 8 can be brought into that position. From a little law that I discovered, I was convinced that they could be folded in the order 1, 5, 6, 4, 8, 7, 3, 2, and also 1, 3, 7, 5, 6, 8, 4, 2, with No. 1 at the top, face upwards, but it puzzled me for some time to discover how. Can the reader so fold them, without, of course, tearing any of the perforation? Try it with a piece of paper well creased like the diagram and number the stamps on both sides for convenience. It is a fascinating puzzle. Do not give it up as impossible!

A diagram similar to Figure F2.1 accompanied the puzzle statement. The eight stamps need only be rectangular, not square.

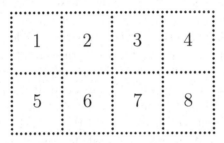

Figure F2.1.   Rectangular block of eight stamps.

The solution to the puzzle appeared a month later (*Strand*, 1920b):

Numbering the stamps as in the diagram last month—that is, 1 2 3 4 in the first row and 5 6 7 8 in the second row, to get the order 1 5 6 4 8 7 3 2, with No. 1, face upwards, only visible, hold this way, with all faces downward: 5 6 7 8.   Fold 7 over 6.
Lay 4 flat on 8 and tuck them both in between $\overset{1\,2\,3\,4}{7}$ and 6, so that these four are in the order, 7 8 4 6. Now bring 5 and 1 under 6, and it is done. The order 1 3 7 5 6 8 4 2 is more difficult and might well have been overlooked if one had not been convinced that, according to law, it must be possible. First fold so that 5 6 7 8 only are visible, with their faces uppermost. Then fold 5 on 6. Now, between 1 and 5 you have to tuck in 7 and 8 so that 7 lies on top of 5 and 8 bends round under 6. Then the order will be as required.

The operative word in the solutions to these two puzzles is "tuck," which implies a bending of some of the stamps as they are slipped into position. If instead of bendable paper stamps we used rigid panels, then there would be no way to solve these puzzles. Thus, these stamp puzzles offer a charming counterpoint to our (relatively) inflexible regimen of flat panels attached by piano hinges.

There are a variety of stamp-folding and related puzzles, which Martin Gardner (1983) described. Such puzzles need not be based on $2 \times n$ blocks and need not involve stamps. Gardner discussed a 3×3 folding puzzle made by R. Lichter of Mt. Vernon, New York, in 1942. The object is to fold the puzzle so that two of the three "boss gangsters," Hitler, Tojo, and Mussolini, find themselves behind bars. Jerry Slocum and Jack Botermans (1986) provided before and after photos of the actual puzzle.

Gardner also presented Robert Edward Neale's "Beelzebub Puzzle." The puzzle involves a 3×3 array, with L Z E in the first row, B B E in the second, and B U E in the third. There are nine puzzles, each folding the block so that all squares are on top of each other and the letters read, from top to bottom, some pseudonym of a fallen angel. These include names such as BEL ZEEBUB and ZEE BUBBLE, and finally perhaps the most difficult: BEELZEBUB. Of course, what makes this puzzle particularly challenging is the multiple occurrence of letters, which gives less of a hint about how to order the squares. It is a safe bet that a corresponding sheet of stamps will not be issued by the United States postal service.

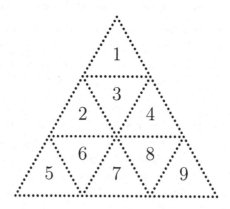

Figure F2.2.   Triangular block of nine equilateral triangular stamps.

Actually, we need not base stamp puzzles on rectangular stamps. Philatelic history contains many examples of triangular stamps, reaching back to 1853 when the Cape of Good Hope first issued them. Indeed, Chris Green (1998) has written an entire book on the subject. Although triangular stamps have come in a variety of different triangular shapes, only three shapes seem suitable for folding puzzles: equilateral triangles, isosceles right triangles, and 60°-right triangles.

In what shape of blocks have equilateral triangle stamps appeared? The largest sheet that I have seen contains the picturesque blue 80-cent 1997 New Zealand Centenary of Pigeon-Gram. The fifty stamps are arranged in five rows of ten stamps each. We can easily extract from the sheet a triangular block that contains nine triangles. Let's number the triangles from 1 to 9, starting with the top row, as shown in Figure F2.2.

Again, we seek to fold the stamps one on top of the other to form a stack. For example, we can produce the order 2 6 7 5 9 3 4 1 8 relatively easily: Fold 3 against 4 and 1 against 4. Fold 1 against 8, then 9 against 3, and then 7 against 9. Finally, fold 6 against 7, and as you do that tuck 5 between 7 and 9.

Try to produce the order 2 5 9 3 4 1 8 7 6, which needs again only one tuck after a number of folds. It is interesting to ask if we can stack the nine stamps so that none of the corner stamps 1, 5, and 9 are next to each other. One ordering for which this holds is the ordering 6 5 4 9 3 2 1 8 7. Again, only one tuck is necessary, after some number of folds.

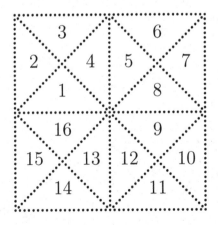

Figure F2.3.   Rectangular block of sixteen isosceles right triangle stamps.

Now let's turn our attention to isosceles right triangle stamps. There is more than one way of laying out these stamps on a sheet. The simplest way, in which each stamp is in one of two orientations, does not seem to be suitable for folding. Another way requires that any point at which the corners of triangles meet should have points all of the same angle, either all 45° or all 90°. An example of the arrangement, containing sixteen stamps, is in Figure F2.3. In 1997 the United States printed sheets of 16 such stamps, labeled "PACIFIC 97" in the denomination of 32 cents. The arrangement has four blocks of four stamps. The sheet was printed in two colors, alternating red stamps with a stagecoach and blue stamps with a clipper ship. In the figure, the odd numbers represent the blue stamps and even numbers the red stamps. As another example, Hong Kong printed the same sheet arrangement of 16 stamps, but of four different denominations, $1.30, $2.50, $3.10, and $5, honoring four of its museums and libraries.

So here is a puzzle for our eager readers: Fold the block of sixteen stamps in Figure F2.3 into a packet sixteen-deep so that you have arranged the stamps in the order 4 1 16 6 5 15 14 8 7 13 11 12 2 3 9 10. More adventurous readers may find additional puzzles for this block of stamps.

As for 60°-right triangles, I know of no stamps produced in that shape. If they had been, then to get foldable blocks we would need sheets in which 30° corners would touch only 30° corners, 60° corners would touch only 60° corners, and 90° corners would touch only 90° corners. I illustrate such a layout in Figure F2.4 and ask readers to fold this block of twelve stamps into a packet twelve-deep with the stamps in the order 5 2 8 9 7 3 4 11 12 1 6 10. Again, adventurous readers may discover additional folding puzzles for this block of stamps.

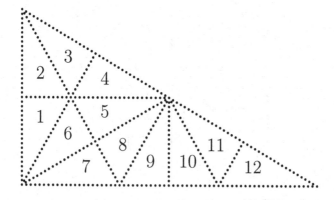

Figure F2.4.   Triangular block of twelve 60°-right triangular stamps.

Chapter 11

Out of the Loop

A beguiling dream for a dissectionist is to find a nifty dissection that some company will market and everyone will buy. In reality there is almost no hope of success, because the set of pieces forms only two different figures, and that won't hold anyone's attention for long. However, as I began to write this book, I discovered that a piano-hinged set of twenty isosceles right triangles could form any of the twelve pentominoes. And what was really great, the piano-hinged pieces formed a symmetrical loop. How lovely!

Unfortunately, when I later searched in a database of U.S. patents for folding objects, I realized that I was not the first to have conjured up this idea. During a time when I had been preoccupied with other projects, I had failed to notice a version entitled "GeoLoop" that the toy company Binary Arts (now ThinkFun) had marketed in the mid-to-late-1990s. That version used a loop of 24 isosceles right triangles to form any of the 35 hexominoes (figures in which six squares are attached edge-to-edge). So, I was really out of the loop, with regard to a puzzle that was made quite simply out of a loop!

Later on, another surprise awaited me. I learned of a loop of sixteen isosceles right triangles that forms a square and all the other tetrominoes (figures from four squares). Entitled "Ivan's Hinge," it was marketed by the puzzle company Paradigm Games, also in the mid-to-late-1990s. So, there is little for me to do but concede that I was doubly out of the loop!

Yet, with the best interests of my readers at heart, I shall explore these remarkable objects and a few variants in this chapter. That way none of you can similarly be thrown for a loop, although I will have of course closed a convenient loophole for you.

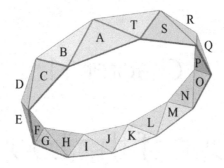

Figure 11.1.   Piano-hinged loop that forms any pentomino.

A pentomino is a figure in which five congruent squares are attached in an edge-to-edge fashion. Not counting reflections, there are twelve different pentominoes. Computer scientist David Eppstein first posed the problem of finding a (swing)-hinged assemblage that could form each of the twelve pentominoes. Erich Friedman refined Eppstein's first attempts to give a lovely 10-piece symmetric dissection, which reduces to eight pieces by merging certain pieces. These are described in a paper by Erik Demaine, Martin Demaine, Eppstein, Friedman, and myself (2005).

Let's see how to accomplish the same for folding dissections. We can form a loop (Figure 11.1) from twenty isosceles right triangles that will fold up to give each of the pentominoes. Figure 11.2 shows four different pentomino foldings, for the X-, W-, Z-, and T-pentominoes, as well as a net for the loop. The perspective views in Figures 11.3–11.6 give a hint as to how these figures form from the folding. We wrap the loop around the pentominoes, in a way reminiscent of mummy wrappings. This is a neat extension of the "gift wrapping" that we saw in Chapter 5. The folding for the X-pentomino is wonderfully symmetric, mimicking the route of a vehicle that keeps looping around a cloverleaf interchange. We can get a 15-piece piano-hinged dissection from the loop, just by gluing pieces O, P, Q, R, S, and T together in the way that they fit together in the four pentominoes of Figure 11.2.

Obviously, this hinging is tube-cyclic. Furthermore, there is a pretty *checkerboarding* property: If the isosceles right triangles alternate between two colors around the loop, then the pentomino squares will alternate in color. Necessarily, the color on one level of each square will be the opposite of the color on the other level.

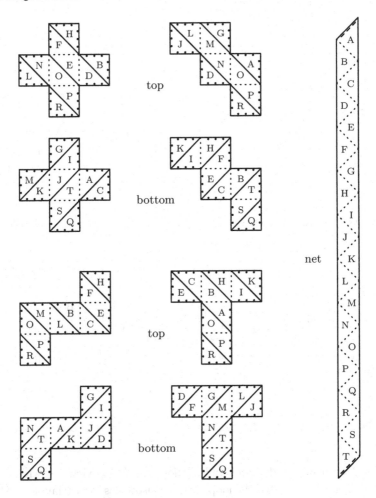

Figure 11.2. Folding dissection of any pentomino to any other pentomino.

Solomon Golomb (1994) has explored the world of polyominoes. Pentominoes are examples of k-ominoes, or polyominoes, for which $k = 5$. In a fashion similar to the above approach, we can form any of the k-ominoes from a loop of $4k$ isosceles right triangles, for any fixed natural number k. Kenneth V. Stevens, an inventor living in Brooklyn, New York, described this loop in his patent (1994). He required that the loop consist of $4k$ isosceles right triangles, but he did not claim that you could form any k-omino from such a loop. It is not hard to prove that claim, using mathematical induction on the value k.

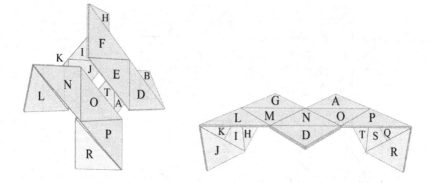

Figure 11.3. Forming an X.   Figure 11.4. Forming a W-pentomino.

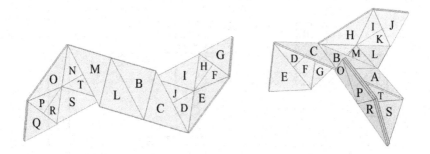

Figure 11.5. Forming a Z-pentomino.   Figure 11.6. Forming a T.

The Israeli puzzle designer and author Ivan Moscovich, together with Jan Essebaggers, patented the loop of sixteen isosceles right triangles (1994). It is not clear whether Moscovich realized that his loop would form all of the tetrominoes, in the same fashion as Stevens's creation. He seemed primarily interested in forming the square, and the patent application also pointed out that a loop of 64 triangles would form a 4×4 square. However, Moscovich marketed it not only as a puzzle but also as a promotional item for a variety of entities, such as the Eastman Kodak Company and the Mid Glamorgan Economic Development Unit in Cardiff, Wales.

Of the twelve pentominoes, the P-pentomino (a 1-square attached to a 2-square) carries a modest surprise with it.

Puzzle 11.1. *Find three distinct ways to form the P-pentomino from the loop of 20 isosceles right triangles.*

I made a model of the loop using triangles of cherry wood that were four inches on each of the two legs and 3/16 of an inch thick. It is easier to manipulate the model when it rests on a flat surface. One can transform one pentomino to another by unfolding along one line and then refolding along another. This technique of successive folds is limited, in the sense that one cannot convert a T-pentomino to a W-pentomino, or an X-pentomino to any other. A more complicated simultaneous motion will carry an X-pentomino to a certain type of P-pentomino, and eventually to other pentominoes. Puzzlists may enjoy cataloging the motions needed to carry one given pentomino to another.

A related puzzle is to dissect any given n-omino and any given m-omino to any given $(m+n)$-omino. Here we can use a variant of the loop approach, since loops in general won't work—just try forming the X-pentomino from a loop for a domino (two attached squares) and a loop for a tromino (three attached squares). Instead, remove one piano hinge from each loop, giving a "ribbon" of triangles. A ribbon for an n-omino can then follow a ribbon from an m-omino, to mimic a ribbon for an $(m + n)$-omino.

Next, choose any natural numbers a and b. For k-ominoes with $k = a^2 + b^2$, we can find a loop of pieces that form any of the k-ominoes and also form a square. The idea is to start with a loop of $4k$ isosceles right triangles and then superpose cuts similar to those that we used in Chapter 10 when we dissected $a^2 + b^2$ equal squares into one large square. The example in Figure 11.7 is for the appropriate set of pieces for $k = 5$, which then form an X-pentomino on the left and a square on the right. Here $a = 1$ and $b = 2$, and we use a total of 32 pieces, which we see in perspective in Figure 11.8

I base the labeling scheme on that in Figure 11.2. An uncut isosceles right triangle gets labeled with the same letter. If I cut an isosceles right triangle, then I cut it into two pieces, one with the same label, and the other with that label underscored. The resulting number of pieces is

$$4(a^2 + b^2 + a + 2b - 2\gcd(a, b)),$$

where $\gcd(a, b)$ is the greatest common divisor of a and b. If we use two colors for the pieces, so that the k-omino is checkerboarded, then the resulting large square will also be checkerboarded. Forsaking checkerboarding, we can save two pieces by gluing together pieces A and T and also pieces D and E. This leaves us with thirty pieces.

All right, all right! This last dissection has left you feeling like you're tied in knots, even if topologists will freely tell you that that's impossible

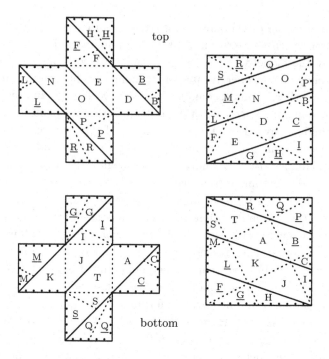

Figure 11.7. Folding of any pentomino to any other pentomino and to a square.

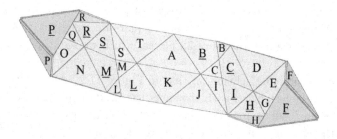

Figure 11.8. Piano-hinged loop well on its way to forming the square.

with a loop. So if these loops make you feel like you're going in circles, take heart! We'll break out of the loop in the next chapter, just wait and see.

Chapter 12

From FIFI to ZULU, and Back!

In the last chapter, we saw a piano-hinged assemblage that folds into a square and into every one of the twelve pentominoes. We paid a high price for that generality: 30 pieces. If we limit ourselves to dissecting just one pentomino at a time to a square, we can do a lot better. Yet, while some of the pentominoes yield without too much trouble, handling the whole lot, from F to Z, isn't so easy. It's tougher than producing the corresponding swing-hinged dissections, which Pieter Torbijn (2001) did. We find ourselves fighting the quirky shapes of these letter-eponymous figures and wondering if it's as difficult as identifying words in that crazy alphabet: *F, I, L, N, P, T, U, V, W, X, Y, Z.* We might expect to meet with *FUTILITY*, but that's not really the case. So let's see what we can piece together, *WILLY-NILLY*, from our *NUTTY* "dodecabet."

The P-pentomino is nothing more than two attached squares, and we have already seen a folding dissection of that figure to a square in Chapter 8. We had better not *PUFF* ourselves up over past glory and run the risk of creating *ILL WILL*. Taking this hint, we proceed on to that most egotistical of all pentominoes, the I-pentomino.

Yes, the I-pentomino seems to stretch out almost to *INFINITY*, which actually helps us. Harry Lindgren (1964b) described a 4-piece dissection that he found by crossposing two strips. Much earlier, Jean Montucla had included this same dissection when he revised the book by Jacques Ozanam (1778), although he did not specifically give the dimensions of the (5 × 1)-rectangle. The dissection is in fact swing-hingeable, although neither Montucla nor Lindgren seemed to realize that. And how could they

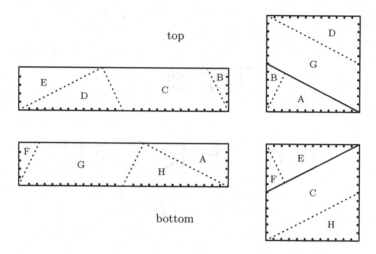

Figure 12.1.   Folding dissection of an I-pentomino to a square.

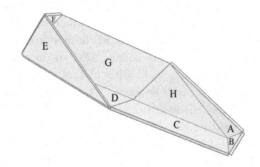

Figure 12.2.   Projected view of an I-pentomino to a square.

INTUIT that there would one day be a lovely symmetric folding dissection based on it?

The folding dissection (Figure 12.1) has just eight pieces, with the levels identical under rotation. Furthermore, we can hinge it so that there are two cap-cycles and one flat-cycle. When we start to transform the I-pentomino to the square by folding, we see a figure reminiscent of a canoe (Figure 12.2). We fold simultaneously along all folds except those between pieces C and D and between pieces G and H, which must stay rigid. Following the simultaneous folding, we can complete the folding along the two piano hinges that we at first left out.

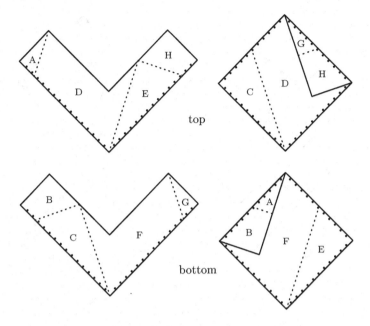

top

bottom

Figure 12.3. Folding dissection of a V-pentomino to a square.

Since the V-pentomino is also symmetric, we look to see if we can find a symmetric dissection of it to a square. Do not *VILIFY* this shape too hastily! We can achieve success with the 8-piece folding dissection in Figure 12.3. The hinging uses one cap-cycle and two flat-cycles, as we can see in Figure 12.4. In folding from the square to the V-pentomino, we must first fold out pieces A and B, and also pieces G and H. Then, all pieces participate simultaneously in the final fold that produces the pentomino.

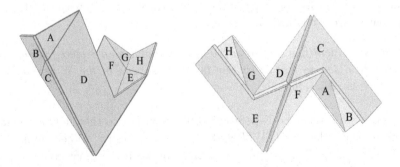

Figure 12.4. V to a square.   Figure 12.5. W-pentomino to a square.

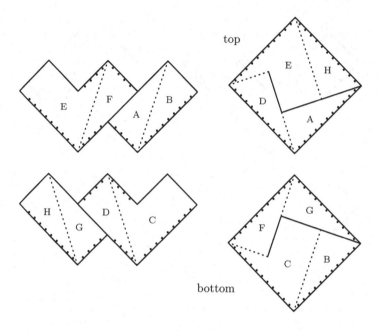

Figure 12.6. Folding dissection of a W-pentomino to a square. (W)

Since the W-pentomino is also symmetric, can we also find a symmetric dissection of it to a square? Let's not *WILT* under the pressure, but just *WILLFULLY* pursue our goal. We *WIN* big time with the *WILY* folding dissection in Figure 12.6. The hinging uses three cap-cycles, as we can see in Figure 12.5. In folding from the square to the W-pentomino, all pieces must move simultaneously, a property that is surely rare. A curious feature of the dissection is that when the pieces form the W-pentomino, the folds that would form the boundary of the square trace out a W, with the strokes alternating from one level to the other.

Puzzle 12.1. *Identify a decomino (10-omino) that has an 8-piece folding dissection to a square that is similar to the dissection of the W-pentomino. Also give the dissection.*

When we attack the N-pentomino, we soon notice that it is not symmetric. Don't be a *NITWIT* and give up too easily. When appropriately viewed, the N-pentomino is a cross between a V-pentomino and a W-pentomino. By suitably blending the dissection technique for the V-pentomino with that for the W-pentomino, we arrive at the *NIFTY* 8-piece

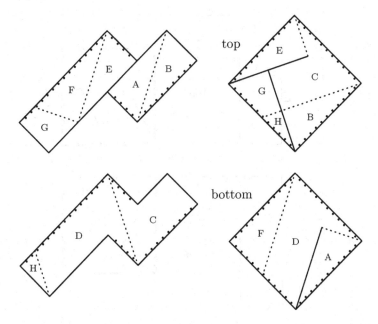

Figure 12.7. Folding dissection of an N-pentomino to a square.

dissection in Figure 12.7. In its cyclic hingings, we get a blend of the V- and W-pentomino characteristics, with two cap-cycles and one flat-cycle, as we can see in Figure 12.8.

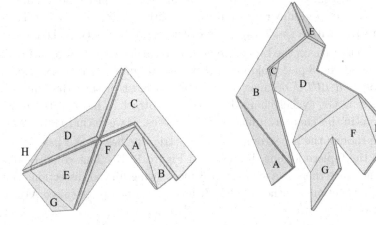

Figure 12.8. N-pentomino to a square.

Figure 12.9. F to a square.

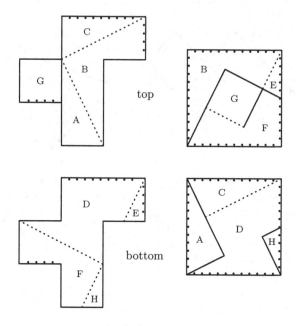

Figure 12.10.   Folding dissection of an F-pentomino to a square. (C)

Named somewhat whimsically, the F-pentomino has a *FUNNY* shape. If we jettison all hopes of symmetry, we can move *FITFULLY* toward an 8-piece folding dissection (Figure 12.10) of the F-pentomino to a square. Again, we see a shape *FULFILL* our expectations: four pieces per level. We also see the barest glimmer of beauty (just a little *FIZZ*—or is it *FLUFF*?) in Figure 12.9 with a single cap-cycle consisting of pieces B, C, D, and E.

Next, let's take on the L-pentomino, from which we hope to get a *LIFT*. I have found the 8-piece folding dissection in Figure 12.11. This folding dissection is a *LULU* because, in contrast to most of the others in this chapter, it is not cyclicly hinged. Also, in contrast to most of the other dissections in this chapter, it has no *TINY* pieces, as we see in Figure 12.12.

Even though the Z-pentomino is symmetric, the best that I have found so far is the 9-piece folding dissection in Figure 12.13. This could be discouraging, since there is a symmetric 3-piece swing-hinged dissection. But still, this dissection is rather *ZIPPY*: It has a Z-shaped piece (piece A) and three cap-cycles, with pieces A, B, C, and D in the first, A, E, F, and G in the second, and A, G, H, and I in the third. When we view the assemblage in perspective in Figure 12.14, we see why we cannot fold out all three cap-

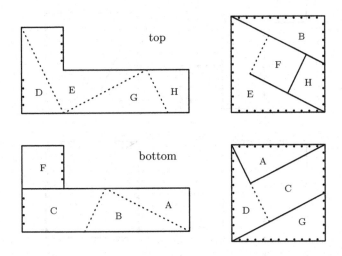

Figure 12.11.   Folding dissection of an L-pentomino to a square.

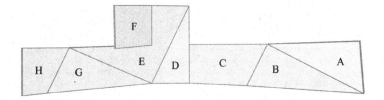

Figure 12.12.   Projected view of an L-pentomino to a square.

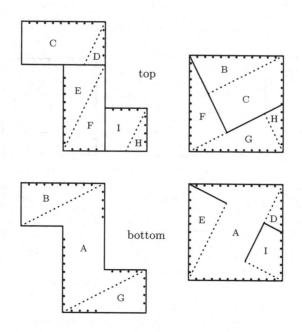

Figure 12.13. Folding dissection of a Z-pentomino to a square.

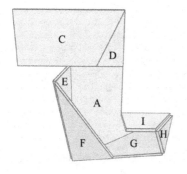

Figure 12.14. Z to a square.

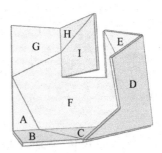

Figure 12.15. U to a square.

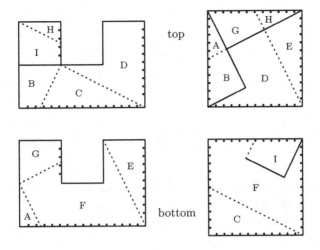

Figure 12.16. Folding dissection of a U-pentomino to a square.

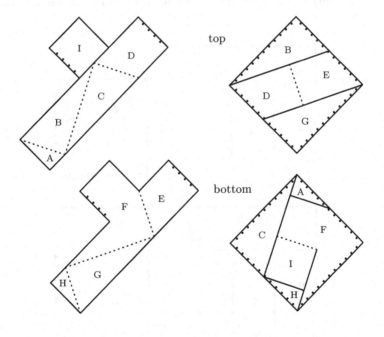

Figure 12.17. Folding dissection of a Y-pentomino to a square. **(C)**

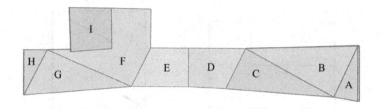

Figure 12.18. Projected view of a Y-pentomino to a square.

cycles at the same time: Pieces E and F get in the way of pieces B (hidden) and C. In transforming the Z to the square, we should first completely fold out pieces E through I. When those are in place in the square, we can then fold out pieces B, C, and D.

Although the U-pentomino itself is quite symmetrical, the symmetry does not seem to impress us with its *UTILITY*. In fact, it was beginning to seem *UNFIT* for any dissection *UNTIL* I found the 9-piece one in Figure 12.16. This hinging has one cap-cycle and two flat-cycles, so it does

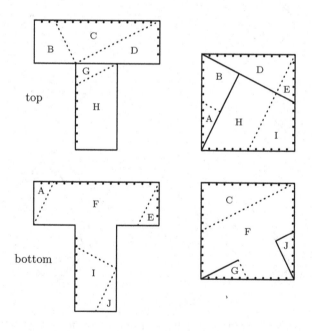

Figure 12.19. Folding dissection of a T-pentomino to a square.

seem to *UNIFY* with other pentominoes. The assemblage confronts us in Figure 12.15.

To find the 9-piece dissection of a Y-pentomino to a square in Figure 12.17, it seems we must do things differently. I positioned the cuts so that the square piece (piece I) that I fold over is not in the middle of the resulting square. In contrast to all of the other pentomino-to-square dissections except the L-pentomino, this dissection is not cyclicly hinged, as we see in Figure 12.18. But nonetheless, it has an intriguing symmetry in the square. *YUP!*

Let's return to one last pentomino that is symmetric: the T-pentomino. We find that symmetry does not always promise nice dissections. Going *FULL TILT*, the best that I have been able to find is the 10-piece folding dissection in Figure 12.19. *TUT-TUT!* Now don't *FLY* into a *TIZZY*, because we still have several cycles: a cap-cycle and two flat-cycles, as we see in Figure 12.20.

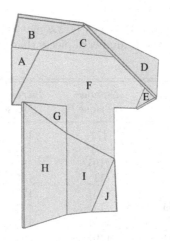

Figure 12.20. Projected view of a T-pentomino to a square.

But what has happened to the X-pentomino, also known as the Greek Cross? Why has it not appeared with all the other pentominoes? Sad to say, no English word over our dodecabet seems to begin with an X. So I have banished the poor letter from this chapter, only to have it sneak back into the book later, disguised as a Greek Cross. Consistent with this circumstance, we have seen no X's in our silly lexicon, though *TWIXT* me and you, we might have been able to *FIX* that.

The Many Sides of Ernest Freese

Ernest Freese challenged himself with many dissection puzzles that had never been tackled before. Nowhere is this more striking than with his dissections of regular polygons to squares. In this regard he was both fearless and inventive, dissecting all regular polygons up to sixteen sides, as well as those of 20 and 24 sides. His audacity and creativity went unmatched for over forty years, until British computer analyst Gavin Theobald produced what may well be the last—and most remarkable—word on these dissections.

We see in Table M4.1 the regular polygons that Freese dissected into squares. I compare the fewest number of pieces that Freese achieved against

Regular polygon	Number of sides	Pieces for Freese	Pieces for Theobald
heptagon	{7}	11*	7
enneagon	{9}	12	9
decagon	{10}	8	7
hendecagon	{11}	13*	10
trikaidecagon	{13}	17*	11
tetrakaidecagon	{14}	14*	10
pentakaidecagon	{15}	17	11
hexakaidecagon	{16}	15	11
heptakaidecagon	{17}	—	12
octakaidecagon	{18}	—	12
enneakaidecagon	{19}	—	15+, 16
icosagon	{20}	19	14
icosikaihenagon	{21}	—	14+, 15
icosikaitetragon	{24}	22	14

* not included in *Geometric Transformations*.
+ allowing pieces to be turned over.

Table M4.1. Dissecting polygons with many sides to squares.

the fewest number that Theobald achieved over four decades later. Note that I do not include entries for the equilateral triangle, the pentagon, the hexagon, the octagon, and the dodecagon, because the best dissections known have not changed since the period during which Freese worked on dissections. Also, for general comparison, I have included the heptakaidecagon, the octakaidecagon, the enneakaidecagon, and the icosikaihenagon even though Freese did not give a dissection for any of them in those materials of his that I have examined.

Theobald (2004) found at least preliminary versions of many of his dissections before I had informed him of Freese's work. While Theobald was having an especially productive period during September of 2003, I mentioned that I was writing this short section and noted that the only many-sided polygon that Freese had handled, but not Theobald, was the {24}. Quickly, and without a peek at Freese's dissection, Theobald found his 14-piece beauty. Theobald went on to reduce the number of pieces of his dissections for {15}, {17}, {18}, and {20} to the number listed in the table, as well as to find his first dissections for {19} and {21}.

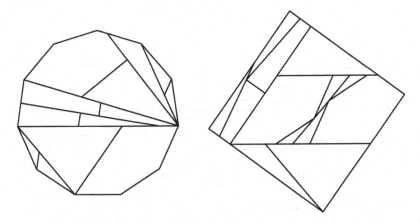

Figure M4.1.   Freese's hendecagon to a square.

Let's start with Freese's 13-piece dissection of a hendecagon to a square (Figure M4.1), which he did not include in *Geometric Transformations*. It is a mystery why he did not include this dissection, as well as several others, in his book. He appears to have been the first person to attempt dissections of the hendecagon and the trikaidecagon, as well as polygons with more sides, into a square. These dissection puzzles are not easy, and Freese's dissections certainly qualify as impressive first attempts.

Freese attacked the hendecagon dissection piecemeal: He cut the figure into seven pieces, six of which together form a parallelogram and the seventh of which is a tall, thin isosceles triangle. He used the P-strip technique to convert the parallelogram into a rectangle of length equal to the desired square. He then used the T-strip technique to convert the isosceles triangle into a second rectangle of length equal to the desired square. We see these two crosspositions in Figure M4.2. Finally, he assembled the two rectangles to give the square.

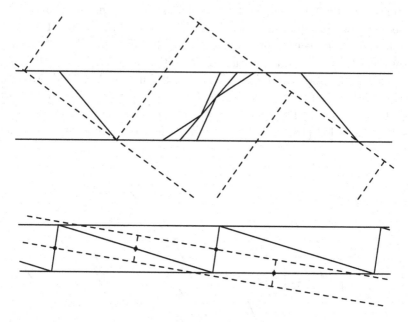

Figure M4.2.   Crosspositions for Freese's hendecagon to a square.

Freese's dissection follows the same basic approach as that of a 9-piece dissection of a heptagon to a square by Harry Lindgren (1964b), which Lindgren (1952) had described in a letter to James Travers. It apparently never occurred to Freese to apply his hendecagon approach to the heptagon, and he settled for an inferior 11-piece heptagon dissection. Indeed, in dissecting the hendecagon, Freese had to also figure out how to slice a long, squat isosceles triangle with an apex of $(9/11)180°$ into two triangles that fit snug against an angle also of $(9/11)180°$. This nifty trick is not needed in forming the parallelogram in Lindgren's heptagon dissection.

Just as Freese's hendecagon dissection was ahead of its time for the 1950s, the next attempt at a hendecagon dissection, a 10-piece beauty by Theobald (Figure M4.3), is remarkable for its time, the early 2000s. Similarly to Freese, Theobald attacked the dissection piecemeal. He cut the hendecagon into four pieces, three of which comprise a T-strip element that has the lion's share of the area. What a delight to see how these pieces fit together into the T-strip element! The remaining piece is by itself a T-strip element, consisting of the merge of an irregular but symmetrical equilateral hexagon with a long, squat isosceles triangle. The resulting odd-shaped piece avoids an extra cut, thus saving a piece in the final dissection.

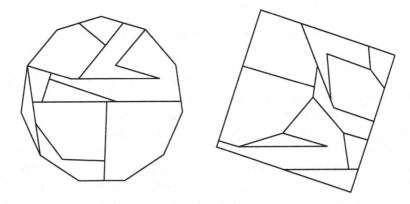

Figure M4.3.   Theobald's hendecagon to a square.

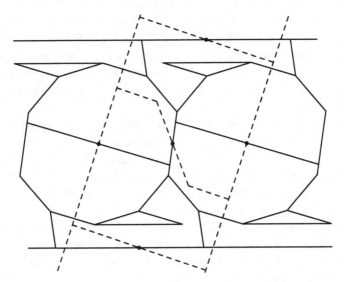

Figure M4.4.   First crossposition for Theobald's hendecagon to a square.

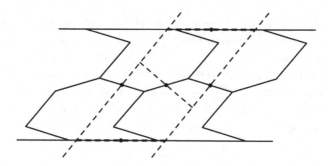

Figure M4.5.   Second crossposition for Theobald's hendecagon to a square.

Rather than produce a pair of rectangles out of the strips, as Freese did, Theobald found an especially convenient trapezoid into which to cut the odd-shaped piece. He then cut the remaining set of three pieces into the shape of a square minus this trapezoid. Once readers have comprehended how Theobald manipulated these T-strips (in Figures M4.4 and M4.5) to produce two figures that fit to form the square, they are ready to ponder how Theobald arrived at the particular choices of the partition of the hendecagon into the four pieces and the two shapes that fit together to make the square. There are an incredibly large number of possibilities that Theobald must have considered in order to identify the final choices that appear so fortuitous. What a superb job he did in topping Freese!

Having considered an impressive dissection that Freese did not even choose to include in his manuscript, along with a remarkable rejoinder from Theobald, we are now ready for even more striking creations. As a second example, we shall compare Freese's dissection of the pentakaidecagon with that of Theobald's.

Freese again approached this dissection piecemeal. He sliced five thin trapezoids off of the exterior of the pentakaidecagon, leaving a regular pentagon. Then, he converted the regular pentagon via a P-strip dissection to a rectangle whose length is that of the desired square. Freese converted the five trapezoids to a thin rectangle whose length is also that of the desired square, using a P-slide, as shown in the accompanying figure. This took eleven pieces, although only ten would be necessary if he had allowed himself to turn over a single piece corresponding to pieces 5 and 6. Altogether, Freese managed the dissection in seventeen pieces, as we see in his Plate 134 (Figure M4.6).

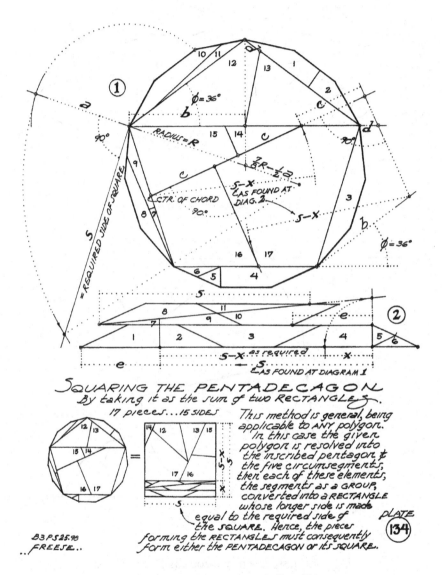

Figure M4.6.   Ernest Freese's Plate 134.

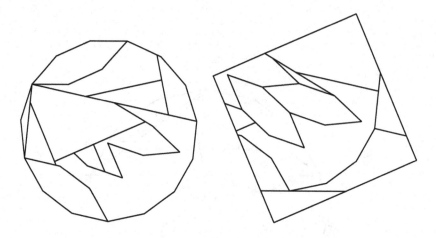

Figure M4.7.   Theobald's {15} to a square.

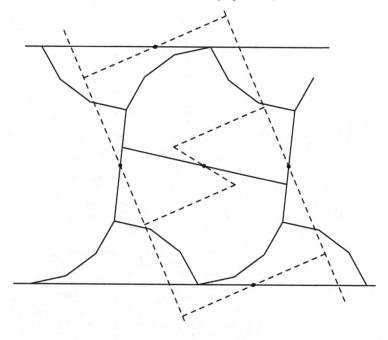

Figure M4.8. First crossposition for Theobald's {15} to a square.

To get his remarkable 11-piece dissection (Figure M4.7), Theobald first partitioned the {15} into five pieces, from which he assembled two T-strip elements. His strategy was to concentrate as much of the area as possible into two large pieces, as we can see in Figure M4.8. Again, the way that these two pieces fit together is mind-boggling! Theobald managed this so that the remaining three pieces, namely two squashed hexagons and an isosceles triangle, form a T-strip element. He could then have crossposed each with a rectangle whose long side would equal the side of the square. Instead, he cut a trapezoidal notch out of the square, choosing the shape of the trapezoid to fit nicely in the crossposition of Figure M4.9. The trapezoid has twice the height of a rectangle for the two squashed hexagons and the triangle.

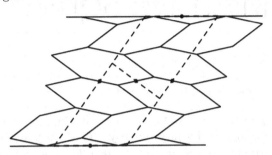

Figure M4.9.   Second crossposition for Theobald's {15} to a square.

By themselves, these are wonderful ideas, which would lead to a 13-piece dissection. Yet Theobald was not done. To reduce the number of pieces (seven) resulting from the second crossposition, he cut a cavity into the largest piece of the square, into which he could insert the two squashed hexagons and the isosceles triangle. He made sure that the relative positions of the hexagons and triangle would conform to the positions of their subpieces when those subpieces fill the trapezoidal notch. This replaces the seven pieces by the five pieces that we see protruding from the upper left corner of the square.

Thus, we have seen Gavin Theobald write miraculous conclusions to the daring stories that Ernest Freese began so many years before. Theobald noted that he benefited from ideas of David Paterson, an Australian civil engineer, who in the 1980s had discovered dissections (unpublished) for many of the puzzles in Table M4.1, though none were as economical as Theobald's. The punch line that Theobald has provided for us deals not so much with one-upsmanship as it does with the sharing of uncommon boldness and imagination. Bravo to all of these geometric explorers!

Chapter 13

Intensive Square

The geometer-on-call logged in her report:

> The polygon was rushed to the Intensive Square Unit after its right-angularity fell precipitously. This followed high-risk surgery attempting simultaneous integrality and perpendicularity, which brought the orthogonality system close to total collapse. With thinking labored, hook-up to an inspirator was indicated, and an emergency procedure to stabilize the angles was performed. While its condition is guarded, the polygon is now responding to therapy. The next 90 degrees will be critical, but if it survives that period, its prognosis should be good, with a complete recovery possible.

Although geometry is an exact science, it is not perfect, as all of you geometers-in-training are aware. Integral identities can lead to dissections of squares that have all of the cuts parallel to the sides. However, it is difficult to predict when we can accomplish this and still save the patient. Let's attempt a few here, and see what our survival rate is.

Any first-year resident will recall from the standard references of Sam Loyd and Henry E. Dudeney a variety of 4-piece unhingeable dissections of squares for $3^2 + 4^2 = 5^2$. More seasoned veterans will be familiar with the recent advances in geometric science that allow us to maintain hingeability for this case, using either my 4-piece swing-hingeable dissection or my 5-piece twist-hingeable dissection. For a piano-hinged dissection, I have developed the 7-piece rounded dissection in Figure 13.1. The main procedure, which will undoubtedly be adopted by ISUs around the world, is to fold piece G down from F along a diagonal fold, changing a primarily vertical piece in the 4-square into a primarily horizontal piece in the 5-square. To

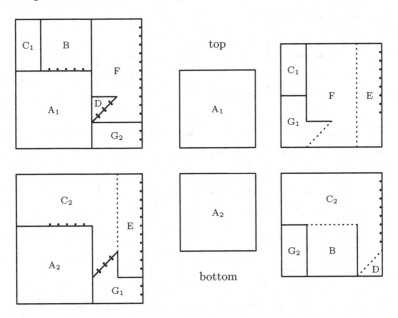

Figure 13.1. Rounded piano-hinged dissection of squares for $3^2 + 4^2 = 5^2$. (C)

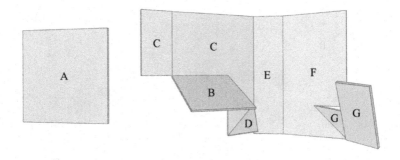

Figure 13.2. Perspective view of squares for $3^2 + 4^2 = 5^2$.

make room next to piece C in the 5-square, and fill in a hole next to piece F, we then fold piece D. We next fold up piece B, and finally fold between pieces E and F. A perspective view is in Figure 13.2.

The simplest dissection of three squares to one is for $1^2 + 2^2 + 2^2 = 3^2$. Indeed, at the dawn of the modern geometric puzzle era, Henry Dudeney (1907) gave three 4-piece dissections for it. Two of them are swing-

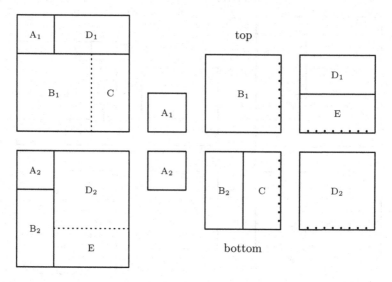

Figure 13.3. Piano-hinged dissection of squares for $1^2 + 2^2 + 2^2 = 3^2$.

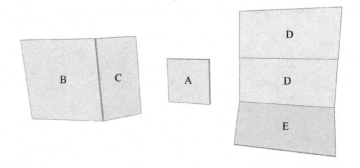

Figure 13.4. Perspective view of squares for $1^2 + 2^2 + 2^2 = 3^2$.

hingeable, but surprisingly, it is the unhingeable one that we can adapt to give the 5-piece piano-hinged dissection in Figure 13.3. I cut the two 2-squares identically and then overlap them in the 3-square. Because of the viewing angle, the complete structure of piece B is hidden in the perspective view of Figure 13.4.

Another dissection of three squares to one is for $4^2 + 4^2 + 7^2 = 9^2$, for which I had found a 6-piece swing-hingeable dissection. My technique in the new piano-hinged dissection is a major advance over the one in Figure 13.3.

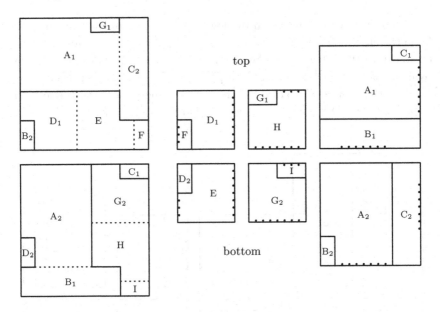

Figure 13.5. Piano-hinged dissection of squares for $4^2 + 4^2 + 7^2 = 9^2$.

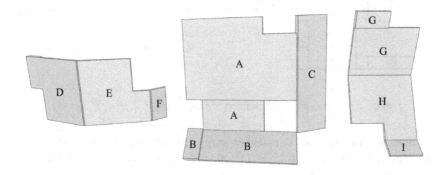

Figure 13.6. Perspective view of squares for $4^2 + 4^2 + 7^2 = 9^2$.

Now we can perform surgery on the 7-square, producing pieces B and C and treating several symptoms simultaneously. When folded out, each piece extends to the boundary of the 9-square on one level and also frees up space on the other level, into which we fit the 4-squares. We can then cut the two 4-squares identically and overlap them in the 9-square.

Pieces F and I extend to fill the lower right corner of the 9-square. To fill in the cavities left by their removal, each of pieces B and C, respectively, expand into the other level. Cavities left by those expansions get filled in by expansions of pieces D and G, respectively, into their other levels. The result is the 9-piece dissection in Figure 13.5. The patient, with one level of pieces C and D hidden in perspective, is recovering in Figure 13.6.

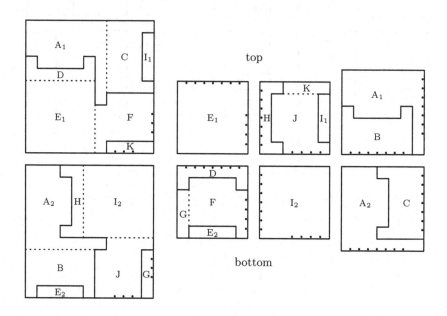

Figure 13.7. Piano-hinged dissection of squares for $6^2 + 6^2 + 7^2 = 11^2$.

How far can we push our newest surgical techniques? Let's see if they are effective on $6^2 + 6^2 + 7^2 = 11^2$, for which Robert Reid had found a 5-piece unhingeable dissection. For piano-hinging, we can follow the lead of the previous dissection. Thus, we can cut the 7-square into pieces A, B, and C, where B and C fold away from piece A and extend out to the boundary of the 11-square. Then, we cut and hinge the two 6-squares identically and overlap them in the 11-square. Furthermore, pieces D and E from one 6-square fill in the area vacated by piece B, while pieces H and I from the other 6-square fill in the area vacated by piece C. The overlap from the two 6-squares consists primarily of pieces F and J.

This basic approach sounds straightforward, but we need dazzling surgical technique to accomplish these objectives: Pieces B, C, D, and H have

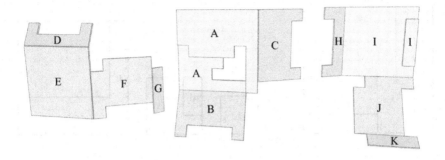

Figure 13.8. Perspective view of squares for $6^2 + 6^2 + 7^2 = 11^2$.

full indentations in the form of a (1×4)-rectangle, and pieces F and J have partial indentations. Into these indentations we will fit (1×4)-rectangles, just like grafting blood vessels in bypass surgery. If you're not knocked out by the anesthesia, you can study the 11-piece dissection in Figure 13.7. Also, be sure to examine the hole in piece A in the perspective view (Figure 13.8).

Your powers of geometric observation have probably alerted you to the fact that in the previous three cases we have delivered twins. Is this a precondition for a successful surgery? Not really, as we see with $2^2 + 3^2 + 6^2 = 7^2$, for which Sam Loyd (*Home*, 1908a) gave a 5-piece unhingeable dissection of squares. Other vital signs include a 6-piece swing-hinged dissection that is hinge-snug. This provides enough encouragement to produce a 10-piece piano-hinged dissection (Figure 13.9). Two of the pieces occupy portions of both levels. These are pieces A and F, as we see in Figure 13.10. Note how the pieces of the 3-square and the 6-square complement each other in the 7-square. Pieces from the 3-square span across the full width of the 7-square. Piece H comes out of a notch in piece F, which is filled by piece C from the 6-square. The technique is effective, but a bit irregular. Can any of you first-year residents find a simpler 10-piece dissection?

If we want our ISU to be truly efficient, we must identify general procedures that allow it to handle whole classes of solutions for three squares to one. As a diagnostic preliminary, we can find all integral solutions to $x^2 + y^2 + z^2 = w^2$ by using a method of the nineteenth-century French mathematician V. A. Lebesgue. Choose m, n, p, and q to be positive integers with $m^2 + n^2 > p^2 + q^2$ and $mq > np$. Then, *Lebesgue's*

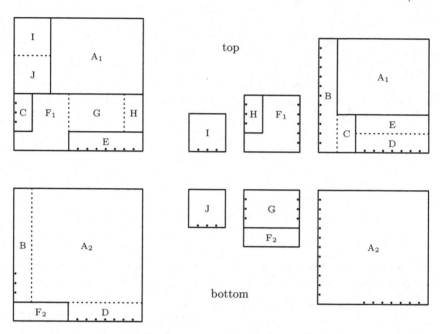

Figure 13.9. Piano-hinged dissection of squares for $2^2 + 3^2 + 6^2 = 7^2$. **(C)**

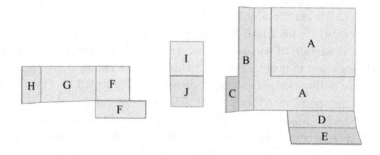

Figure 13.10. Perspective view of squares for $2^2 + 3^2 + 6^2 = 7^2$.

formula is

$$x = m^2 + n^2 - p^2 - q^2, \qquad y = 2(mp + nq), \qquad z = 2(mq - np),$$
$$w = m^2 + n^2 + p^2 + q^2.$$

We can then wheel into our surgical theater a whole class of integer solutions to $x^2 + y^2 + z^2 = w^2$ that I have called the *square-sum-minus*

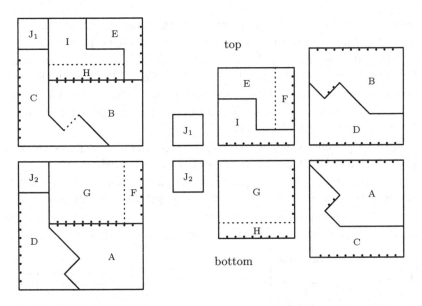

Figure 13.11. Rounded piano-hinged squares for $16^2 + 50^2 + 40^2 = 66^2$.

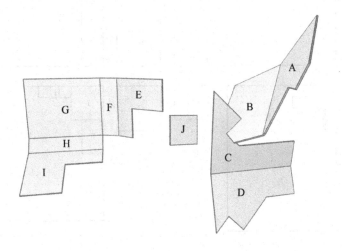

Figure 13.12. Perspective view of squares for $16^2 + 50^2 + 40^2 = 66^2$.

class. This patient population results from restricting $p = 0$, $q = n$, and $m = n - 1$ in Lebesgue's formula, ensuring that $x + y = w$. For example, patient $n = 5$ will have $16^2 + 50^2 + 40^2 = 66^2$. As soon as the anesthesia kicks in, perform a flap-step conversion of the 40-square to a (50×32)-rectangle. Then, transform the 50-square to an L-shape by converting a twist-hinged dissection via the procedure from Chapter 4. The resulting 10-piece rounded piano-hinged dissection is in Figure 13.11, with the perspective in Figure 13.12. Two of the piano hinges abut in the 66-square, but this is not really a problem. We need only spin the (50×32)-rectangle by $180°$ to separate the piano hinges.

For any patient n, we will convert the z-square via a flap-step to a $(y \times 2x)$-rectangle. Patient $n = 3$ (with $4^2 + 18^2 + 12^2 = 22^2$) gets the same treatment. Almost every patient n will pay the same 10-piece charge. Fortuitously, patient $n = 2$ (with $1^2 + 8^2 + 4^2 = 9^2$) gets a special rate of nine pieces. This is because the piano-hinged dissection of an $(a \times 2b)$-rectangle to a $(2a \times b)$-rectangle needs just four pieces, rather than five.

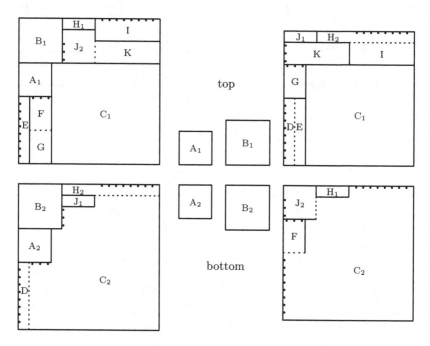

Figure 13.13. Piano-hinged dissection of squares for $3^2 + 4^2 + 12^2 = 13^2$.

With a facility to successfully handle a whole class, our ISU has suddenly accumulated considerable clout. When we perfect a way to handle a second whole class, we demand, and get, a second dissecting theater. We name it after the eighteenth-century Italian mathematician Pietro Cos-

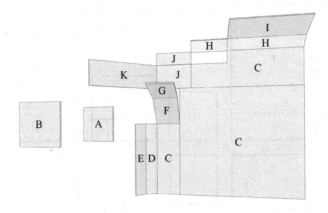

Figure 13.14.   Perspective view of squares for $3^2 + 4^2 + 12^2 = 13^2$.

sali, who identified a class of identities that includes $3^2 + 6^2 + 2^2 = 7^2$, $3^2 + 12^2 + 4^2 = 13^2$, and $5^2 + 20^2 + 4^2 = 21^2$. It results from taking $m = p + 1$, $n = m$ or $n = p$, and $q = n$ in Lebesgue's formula. Our general procedure for this class comes from converting the R-flaps of Figure 8.17 to flap-steps. When making the conversion, one flap-step will use absorption and the other will not. The result is an 11-piece piano-hinged dissection. Figure 13.13 illustrates the method for $3^2 + 12^2 + 4^2 = 13^2$, where $n = 2$ and $p = 1$. As we see in Figure 13.14, the flap-step consisting of the four pieces D, E, F, and G, as well as the left part of piece C, does not use absorption, while the flap-step consisting of the four pieces H, I, J, and K, as well as the top part of piece C, does use absorption.

Now that our ISU is well-established, we should not be surprised to encounter the unusual set of squares. Ernest Freese (1957b) identified a 5-piece unhingeable dissection of squares for $9^2 + 12^2 + 20^2 = 25^2$. That comes easily by converting the 12-square to a (9×16)-rectangle using a simple step dissection, and similarly converting the 20-square to a (16×15)-rectangle. For a rounded piano-hinged dissection, we can use the flap-step to perform the conversions. The resulting 11-piece dissection is in Figure 13.15, and a perspective view is in Figure 13.16.

Having done a great job on that last referral, we might anticipate further referrals from the same source. Ernest Freese also identified a 6-piece unhinged dissection of squares for $2^2 + 4^2 + 5^2 + 6^2 = 9^2$. There is a simple 10-piece piano-hinged dissection for it. Some cases are so easy that even the

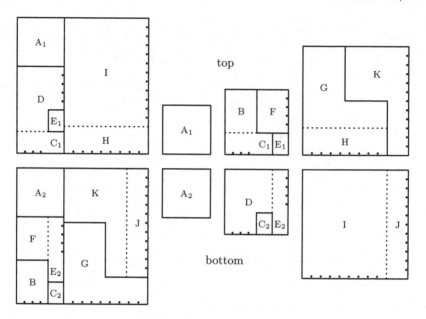

Figure 13.15. Rounded piano-hinged squares for $9^2 + 12^2 + 20^2 = 25^2$.

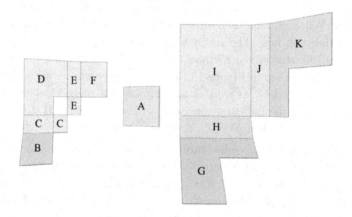

Figure 13.16. Perspective view of squares for $9^2 + 12^2 + 20^2 = 25^2$.

geometers-in-training are allowed to work on them unsupervised. Finish up this one, and you should be ready to be board-certified.

Puzzle 13.1. *Find a 10-piece piano-hinged dissection for squares for* $2^2 + 4^2 + 5^2 + 6^2 = 9^2$.

Chapter 14

Staying Rational

When we shift our attention away from squares to other regular polygons, it may at first seem traumatic to find ways to deal with integer identities. But do not despair! There are treatments that have demonstrated their effectiveness and from which you will derive positive benefit. In fact, as you will see, they may even help preserve your sanity. Then, you will be able to repeat, just like Chief Inspector Dreyfuss, shortly before his relapse, "Every day, and in every way, I am getting better and better." Yes, you will develop the mental toughness to handle triangular relationships. And you will have made such progress by the end of the chapter that you will not hesitate to confront a few latent hexagons.

The first identity, $1^2 + 2^2 + 2^2 = 3^2$, seems almost too easy, but it's perfect to use as a confidence-builder. There is a simple 4-piece swing-hinged dissection of triangles for $1^2 + 2^2 + 2^2 = 3^2$ and also a 4-piece twist-hinged dissection. Of course, the latter converts to a 6-piece piano-hinged dissection, but we can do better than that with the elegant 5-piece piano-hinged dissection in Figure 14.1. In a manner similar to the piano-hinged dissection of squares for $1^2 + 2^2 + 2^2 = 3^2$ in Figure 13.3, cut the two 2-triangles identically, then overlap them in the 3-triangle. We see in Figure 14.2 that the assemblage for one of the 2-triangles is really the same as the other, but just rotated 180°.

Isn't it reassuring that triangles, like squares, can keep their composure for a piano-hinged dissection with $1^2 + 2^2 + 2^2 = 3^2$? Since squares also hold up so admirably under $4^2 + 4^2 + 7^2 = 9^2$, we are encouraged to hope that triangles may maintain their equilibrium in this case too. And they do, as demonstrated by my 9-piece rounded piano-hinged dissection in Figure 14.3. As with the piano-hinged dissection of squares for $4^2 + 4^2 + 7^2 =$

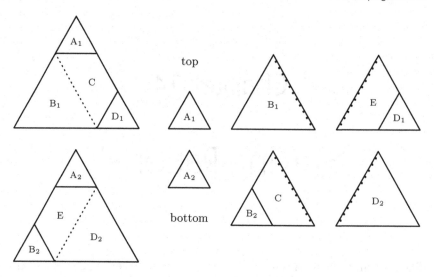

Figure 14.1. Piano-hinged dissection of triangles for $1^2 + 2^2 + 2^2 = 3^2$.

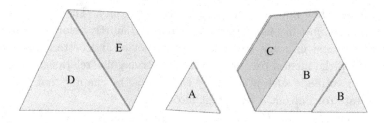

Figure 14.2. Projected view of triangles for $1^2 + 2^2 + 2^2 = 3^2$.

9^2 in Figure 13.5, we cut and fold out pieces from the 7-triangle. Pieces A, B, D, and E extend down to help form the base of the 9-triangle. We then cut the two 4-triangles identically, producing pieces G and I, which fill in the cavities evacuated by pieces B and D, respectively. It is gratifying to see the pieces from the three smaller triangles cooperate so harmoniously in the 9-triangle. Note in Figure 14.4 that pieces G and I from the two 4-triangles are poised for overlap in the hexagonal center of the 9-triangle.

Now that triangles have survived their episode with $4^2 + 4^2 + 7^2 = 9^2$, will they crack under the pressure of $6^2 + 6^2 + 7^2 = 11^2$? The answer is in my 9-piece rounded piano-hinged dissection in Figure 14.5. We first cut the two 6-triangles identically, so that we can fold and overlap them in the

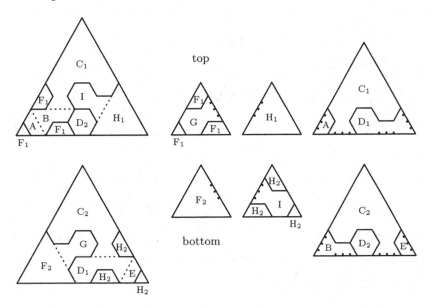

Figure 14.3. Rounded piano-hinged dissection of triangles for $4^2 + 4^2 + 7^2 = 9^2$.

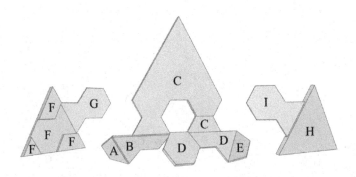

Figure 14.4. Projected view of triangles for $4^2 + 4^2 + 7^2 = 9^2$.

11-triangle, neatly filling out the lower part of that triangle. We then cut and fold down pieces B and C from the 7-triangle to fill in the cavities left after folding out pieces E and H in the 6-triangles. At the same time, pieces E and F, and H and I, fill the cavities left in piece A after folding down pieces B and C. The lovely, soothing symmetry shows up clearly in Figure 14.6.

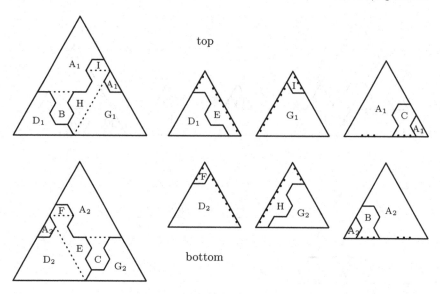

Figure 14.5. Rounded piano-hinged triangles for $6^2 + 6^2 + 7^2 = 11^2$. (C)

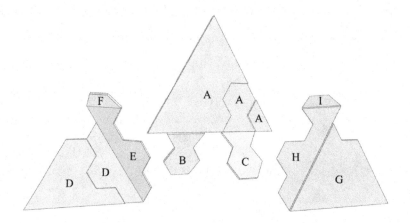

Figure 14.6. Projected view of triangles for $6^2 + 6^2 + 7^2 = 11^2$.

Yes, yes, ... we know that one of the triangles in each of those dissections has an "evil twin." If you will only calm down, you will see that there is no cause for such anxiety. Take a close look at the triangles for $2^2 + 4^2 + 5^2 + 6^2 = 9^2$ in Figure 14.7. See, there is no evil twin controlling the very being

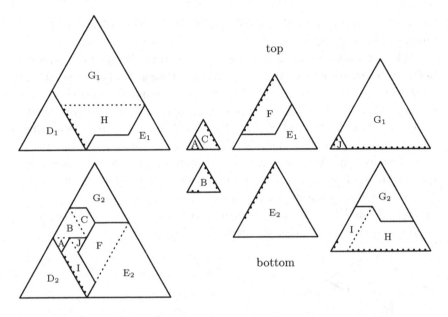

Figure 14.7. Piano-hinged dissection of triangles for $2^2 + 4^2 + 5^2 + 6^2 = 9^2$.

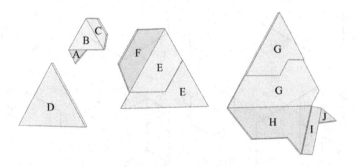

Figure 14.8. Perspective view of triangles for $2^2 + 4^2 + 5^2 + 6^2 = 9^2$.

of the other. Observe further that there are only ten pieces. And the 4-triangle looks so peaceful; it is left completely alone, even in the perspective view (Figure 14.8). See how the 5-triangle and 6-triangle get along so well together, each folding a flap down (pieces F and H, respectively) to give the other space. What, you don't understand why pieces I and J are doubled over, and why the 4-triangle has disappeared, and why the 2-triangle seems to be having a breakdown? Can't you see, it's all for the better? Yes,

I'm sure it could seem a bit confusing and, ah, a bit disorienting—Nurse! Orderly! Come quickly!

This last identity has pushed our triangles to the brink. Let's back off before we suffer a total collapse and shift our attention to hexagons. There are several good reasons why $1^2 + 2^2 + 2^2 = 3^2$ may not be a shock to their psyches. First, there is a simple 6-piece unhingeable dissection of hexagons for $1^2 + 2^2 + 2^2 = 3^2$. Second, Robert Reid found an attractive 7-piece swing-hinged dissection for these figures. Finally, I have found an 8-piece twist-hinged dissection. Yet, the prognosis is more guarded for piano-hinging these figures. The most promising approach is to conduct an inventory of our strengths and thus find a network of support and inspiration in Figures 13.3 and 14.1. Combined with perseverance, they lead to the nifty 11-piece rounded piano-hinged dissection in Figure 14.9.

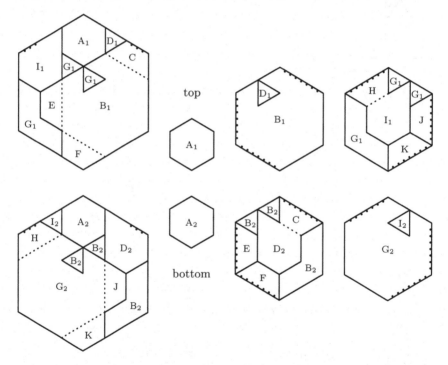

Figure 14.9. Rounded piano-hinged hexagons for $1^2 + 2^2 + 2^2 = 3^2$. (C)

The 1-hexagon fills in the upper corner of the 3-hexagon, while most of the top level of one 2-hexagon fills in the lower right corner of the top level of the 3-hexagon and most of the bottom of the other 2-hexagon fills

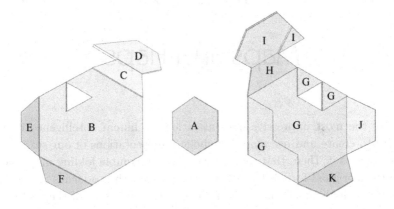

Figure 14.10. Projected view of hexagons for $1^2 + 2^2 + 2^2 = 3^2$.

in the lower left corner of the bottom level of the 3-hexagon. We can then fold out pieces of each 2-hexagon to fill portions of the 3-hexagon. The triangular towers on pieces B, D, G, and I are the result of a swapping around that saves two pieces. This swapping around produces the holes in pieces B and G that we see clearly in Figure 14.10. If we turn over one of the 2-hexagons relative to the other, we see that I have cut and piano-hinged them identically. Now doesn't that make you feel that all is right in the world?

Oh, no! There he goes again, ranting on about an evil twin, who this time even has an identical hole in the head! Why are our polygons so jealous of each other? Why can't they deal with such beautiful symmetry? They'll drive us all crazy yet! Thank goodness, our break is coming up now.

Folderol 3

Maps and Flaps

One of the most impressive demonstrations of human intelligence is the ability to create and use maps, symbolic representations of our surroundings. Why is it, then, that we all have so much trouble folding maps back into their original form? The folds that give us such headaches are generally simple folds, made down the whole length or breadth of the packet as we compress the map. It is easy to unfold the map, and indeed it is not so hard to refold the map if no one has disrupted the pattern of mountain and valley folds by a hasty, bungling attempt at refolding. Yet, these sloppy misfolds induce more false starts, which leave the map in even worse shape.

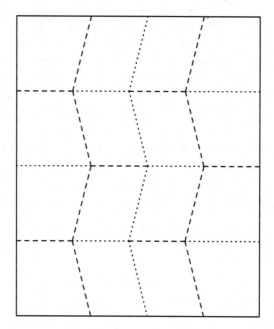

Figure F3.1. Miura-Sakamaki-Suzuki map folding.

Koryo Miura, Masamori Sakamaki, and K. Suzuki (1980) proposed a radically different method for folding a map. Their approach avoids the

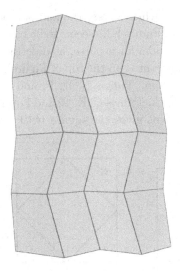 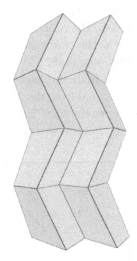

Figure F3.2.   Starting to fold the map.   Figure F3.3.   Folding it further.

simple folds that get us into trouble. Instead, the folds zigzag across the surface of the map in a way that allows the map to open like an accordion just by pulling apart two opposite tabs. Folding the map back up is simply a matter of squooshing the sections back together. It's surprisingly easy!

The hard part is to place the system of folds in the paper to begin with. Figure F3.1 gives a sample of the system. I indicate each convex fold, or *mountain fold*, by a dashed edge and each concave fold, or *valley fold*, by a dotted edge. Starting with the map folded flat as in Figure F3.1, we see the map partially folded together in Figure F3.2, and then we see the map begin to buckle in Figure F3.3. Eventually, all sixteen panels will lie stacked one on top of the other.

This map-folding technique is an example from the field of *flat origami*, in which the resulting folded figure lies flat against itself. Although the main topic of this book is similar in spirit to flat origami, the model is necessarily somewhat different. Origami often has several levels of paper folded along the same crease line. In addition to folding along the crease line, the paper, because it has "body" to it, must also bend and stretch near the crease line. To avoid dealing with this bending and stretching, a mathematical model of flat origami assumes that the paper has no thickness, an assumption that we explicitly avoided with our panels in Chapter 2.

Flat origami is related to origami tessellations, which Helena Verrill has produced in remarkable variety. Let's examine a simple component of

origami tessellations, the *square twist*, for which we see the fold pattern in
Figure F3.4. The resulting square twist in Figure F3.5 seems to be basically
the same as the traditional (Japanese) model for a purse, as described by
Shuzo Fujimoto (1982). A view from the front is on the top of Figure F3.5,
and a view from the back is on the bottom. Panels B, D, F, and H are
totally obscured from the front and the back. Panels C, E, G, and I overlap,
as is clear in the view from the back. This gives the square twist a nice
locking mechanism—perfect for a purse!

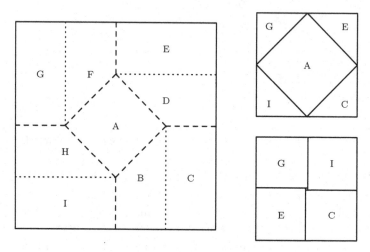

Figure F3.4. Fold pattern for the square twist.   Figure F3.5. Front and back.

To either fold the square twist or to unfold it, it seems to be necessary
to bend the paper, even in the model in which the paper has no thickness.
Suppose that we are not interested in having a purse that locks, but would
rather have a surface that unfolds easily and is less subject to wear and tear.
Can we introduce additional folds into the origami so that we do not bend
any of the remaining surfaces of the paper? The origami in Figure F3.6
does just this. The resulting modified square twist is in Figure F3.7. The
additional folds seem to be precisely what is needed to relieve the pressure
that causes the bending in the traditional square twist. We see the modified
square twist beginning to fold in Figure F3.8, and see it approximately half
of the way to its goal in Figure F3.9.

Can we remove the self-locking mechanisms from other more complicated origami designs? From the viewpoint of traditional origami, this may
be the wrong question to ask, but some traditional origamists (origamis-

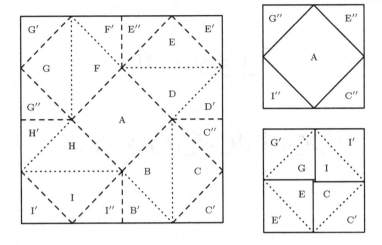

Figure F3.6. Pattern: modified square twist.   Figure F3.7. Front and back.

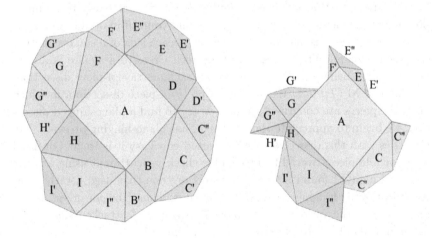

Figure F3.8. Starting the modified twist.   Figure F3.9. Folding it further.

tas?) have thought about *rigid origami.* Thomas Hull (2003) summarized the current state of such work. He even described a different mountain-valley assignment to the folds in Figure F3.4 that made the folding rigid, although face A no longer lands in the front. So, challenges still abound in this area!

Chapter 15

Double-Crossed!!

Now that we've become adept at piano-hinged dissections, it's time to bare—or is it bear?—our crosses. So lovely and symmetrical, these innocent objects seem full of promise. Yet, when we go to two levels, they throw double trouble our way. In exasperation, you may well declare that you have been the victim of a classic double-cross! So, get ready to mount a double-barreled assault, or you could well find yourself in double jeopardy!

A *Greek Cross* results when we glue on to each side of a square another congruent square. The dissection of a Greek Cross to a square is an old favorite. Don Lemon (1890) published an asymmetrical 4-piece dissection, and Sam Loyd (*Tit-bits*, 1897a) gave a pretty 4-piece dissection, in which all of the pieces are congruent. When we try to find a piano-hinged dissection, we begin to appreciate the challenge that piano-hinging presents. My first tries with this dissection demonstrated how easy it is to get confused when we fold pieces over. I had to double-check each attempt using a paper model. Finally, I found a 12-piece dissection, one that used a rather unusual leaf-cyclic hinging. Almost a year later, when I began investigating dissections of the various pentominoes to a square, I attacked this puzzle again and found the 10-piece dissection in Figure 15.1.

The hinged assemblage has a lovely rotational symmetry. On the top level of the cross, we cut off its right arm, and on the bottom level we cut off its left arm. Next, we fold over the rest of the top level, bringing pieces D and E into the same plane as pieces F and G. Then, we fold along the line separating pieces D and F from E and G, all four of which constitute a flat-cycle. In addition, we see two cap-cycles in Figure 15.2, each folded halfway from their positions in the cross to their positions in the square.

The Greek Cross is a k-omino, where $k = a^2 + b^2$, and specifically $a = 1$ and $b = 2$. We could adapt from Chapter 11 the very general dissection of

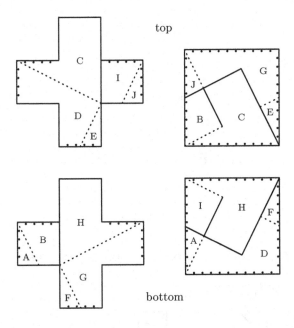

Figure 15.1. Folding dissection of a Greek Cross to a square. (W)

any such k-omino to any other such k-omino or a square. However, that approach requires more pieces. If we fold the loop to fill in the Greek Cross, combining isosceles right triangles along edges not on the boundary of the Greek Cross, we find that we use twelve pieces, six on each level. In the square, we must fold the loop eight more times. This gives a total of twenty pieces, double the number that we have used in Figure 15.1.

Henry Dudeney (*Dispatch*, 1900a) gave a lovely 5-piece dissection of two Greek Crosses to one. The dissection is even swing-hinged, but it is not hinge-snug. Consequently, we cannot convert it to be twist-hinged and then further convert it to be folding. There is a 6-piece swing-hingeable dissection based on the 5-piece dissection that is hinge-snug. That converts to a 10-piece twist-hinged dissection, which seems difficult to convert to a 20-piece piano-hinged dissection, because several of its pieces are decidedly nonconvex. It seems easier to find an 8-piece dissection that we can adapt into a folding dissection. The resulting 20-piece dissection is in Figure 15.3.

I cut and hinge each of the small Greek Crosses identically, resulting in rotational symmetry for the large Greek Cross. The crosses also have rotational symmetry, rotating the top level into the bottom. We can hinge

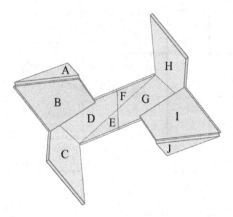

Figure 15.2. Projected view of a Greek Cross to a square.

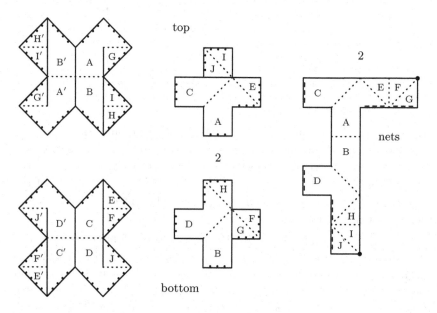

Figure 15.3. Folding dissection of two Greek Crosses to one. **(W)**

each small cross with three cycles, producing a lovely mechanism to flex from one figure to the other, as we can see in Figure 15.4. Two of these are cap-cycles, and the third is a tube-cycle. To produce the net, we must break each of these cycles, as indicated by dashes in the net. Also, as we

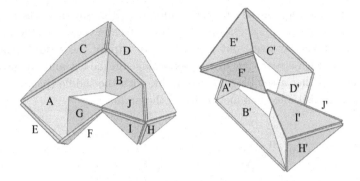

Figure 15.4. View of the assemblages for two Greek Crosses to one.

flex each of the small Greek Crosses, we can keep a vertex shared by pieces F and G adjacent to a vertex shared by pieces I and J. These vertices are marked with small dots in the net. This dissection is exterior-preserving.

I hope that you are enjoying this dissection, because it is actually part of a double feature. Remarkably, with the same set of pieces but a different

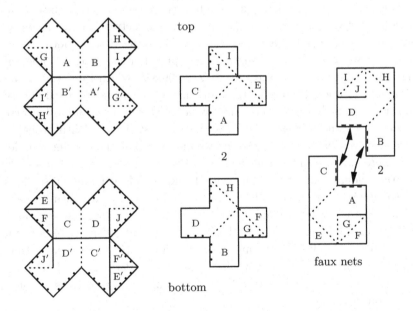

Figure 15.5. Second folding dissection of two Greek Crosses to one. (W)

hinging, we can replace the tube-cyclic hinging with a saddle-cyclic hinging (Figure 15.5). In the process, we lose the four cap-cycles, but the hinged dissection gains the inside-out property. We admire the saddle cycles in Figure 15.6, with a rear view of one assemblage on the left and a view from above of the other assemblage on the right, with the saddle (pieces A′, B′, C′, and D′) positioned symmetrically in the middle.

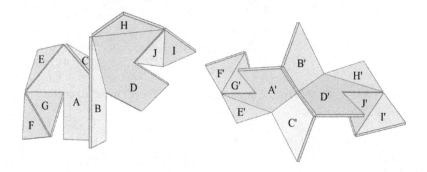

Figure 15.6.   View of saddle-cyclic two Greek Crosses to one.

It is not hard to see that the angle between pieces A and B must equal the angle between pieces C and D, and the angle between pieces A and C must equal the angle between pieces B and D. In the views of both assemblages, I have chosen all four angles to be equal. For this case, the dihedral angles must be $180° - \arccos(3 - 2\sqrt{2}) \approx 99.88°$, which I determined by using vector products—my first use of that operation in 35 years! Those who have studied trigonometry may recall that the *arccosine*, or *inverse cosine*, of a value r is the angle α such that $\cos(\alpha) = r$. With probably more than you bargained for, it's no wonder if you are now seeing double!

A *Latin Cross* results when you form a Greek Cross and then glue a sixth congruent square onto its bottom arm. Having found the folding dissections of two Greek Crosses to one, I was encouraged to try for two Latin Crosses to one. Searching for a piano-hinged dissection of a pair of these crosses puts us in a double bind. The sixth small square that differentiates a Latin Cross from a Greek Cross seems to destroy any hope of symmetry. How can we fill out the long arm in the large cross? Persevering, I recycled some of the tricks that I used in Figures 15.3 and 15.5 to get the 22-piece dissection in Figure 15.7.

The small cross on the left has a saddle-cyclic hinging, and the small cross on the right has a cap-cyclic hinging. To produce the net and faux net,

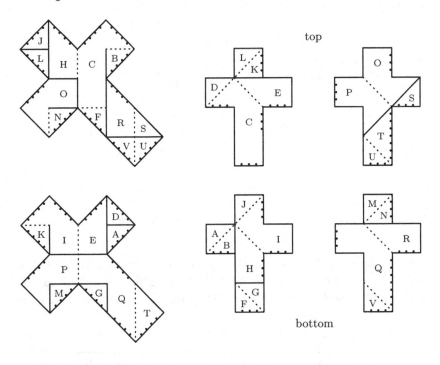

Figure 15.7. Folding dissection of two Latin Crosses to one.

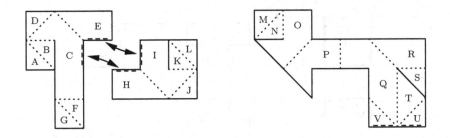

Figure 15.8. Net and faux net for folding dissection of two Latin Crosses to one.

we must break each of these cycles, as indicated by dashes in Figure 15.8. The small cross on the left has the inside-out property, while the cross on the right has the exterior-preserving property. Perspective views of the assemblages are in Figure 15.9. Again, I have displayed the saddle-cyclic hinging with equal angles, which are identical to those in Figure 15.6.

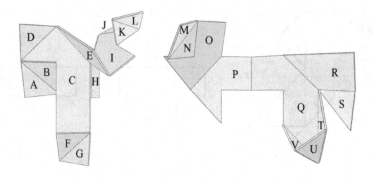

Figure 15.9. View of two Latin Crosses to one.

It is easy to find a line segment of length $\sqrt{5}$ inside a Greek Cross. Thus, it is not difficult to find an 9-piece swing-hingeable dissection of Greek Crosses for $1^2 + 2^2 = (\sqrt{5})^2$. A flappable dissection is a bit more

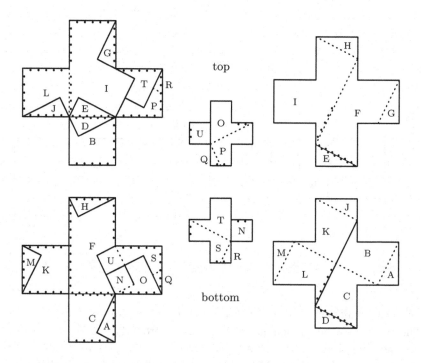

Figure 15.10. Folding dissection of Greek Crosses for $1^2 + 2^2 = (\sqrt{5})^2$.

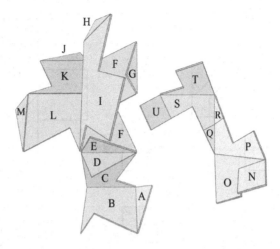

Figure 15.11.   View of Greek Crosses for $1^2 + 2^2 = (\sqrt{5})^2$.

challenging. A workable plan is to dissect the 2-cross into four fifths of the $\sqrt{5}$-cross, all except for the right arm. Then, use the flappable dissection in Figure 15.1 to convert the 1-cross to the remaining square of the $\sqrt{5}$-cross.

To dissect the 2-cross, we can use the leaf-cyclic strategy discarded from my earlier 12-piece dissection of a Greek Cross to a square. Pieces F and I in Figure 15.10 fold together to give the main portions of one arm and the center of the $\sqrt{5}$-cross. Pieces K and L fold to give the main portion of the left arm, and pieces B and C give the bottom arm. This gives fourteen pieces for the 2-cross, plus ten pieces for the 1-cross. But this is only the first part of a double play, as a close inspection of Figure 15.10 reveals. We can save two pieces from the 1-cross and one from the 2-cross, if we do not fold piece I. This allows us to get by with 21 pieces rather than 24. Just to cross you up: While the 2-cross has the inside-out property, I have specified the folds in the 1-cross to be exterior-preserving, as you can see in Figure 15.11.

I like to call the next dissection a double-decker, because there are so many pieces that extend into two levels. It animates the special relationship $1^2 + (\sqrt{8})^2 = 3^2$. Finding a flappable dissection for this relationship is again a bit more challenging, but I eventually produced the 18-piece dissection in Figure 15.12. Let's start by dissecting the $\sqrt{8}$-cross into most of the 3-cross. Creating a cap-cycle for pieces D, E, F, and G suggests extending pieces D and G into both levels. We can then find a cap-cycle for pieces

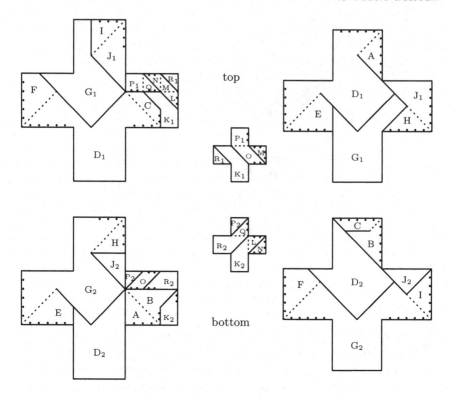

Figure 15.12. Piano-hinged dissection of Greek Crosses for $1^2 + (\sqrt{8})^2 = 3^2$.

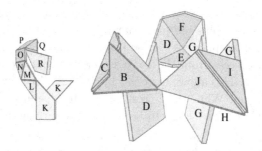

Figure 15.13. View of Greek Crosses for $1^2 + (\sqrt{8})^2 = 3^2$.

A, B, C, and D, as we see in Figure 15.13. There is also a cap-cycle for pieces G, H, I, and J, which is surprising considering that piece J extends into both levels. A remarkable feature is that pieces B and J touch in the three configurations of Figures 15.12 and 15.13. If we connect these two pieces with a universal joint, will we still be able to fold from one cross to the other?

If we make pieces B and C isosceles right triangles, then this would leave a hole in the form of an L-pentomino, into which we must fit a Greek Cross. Of course, we could do this in 15 pieces, by using the loop that we studied in Chapter 11. We can reduce this to nine pieces if we glue additional triangles together. However, we can do a bit better by trading a small triangle from piece C to piece B, as I have already done in Figure 15.12. We can fill the remaining shape in the 3-cross by splitting the 1-cross into just eight pieces. To fill in the details, you may want to at first ignore this last optimization:

Puzzle 15.1. *Find a 9-piece piano-hinged dissection of Greek Cross to an L-pentomino.*

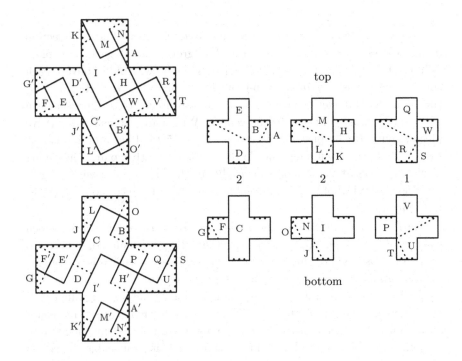

Figure 15.14. Folding dissection of five Greek Crosses to one.

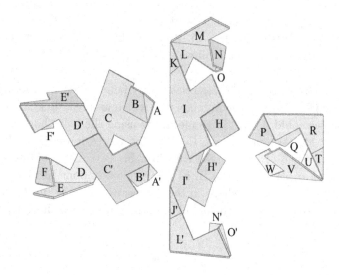

Figure 15.15.   Halfway-folded five Greek Crosses to one.

Of course, we can use the $\sqrt{5}$ in the Greek Cross to find a dissection of five Greek Crosses to one. Harry Lindgren (1964b) gave a 12-piece unhingeable dissection. For a folding dissection, we could convert each small Greek Cross to a square, using the dissection in Figure 15.1. However, this is clearly inefficient, since we do not need to make folds where one arm of the large cross connects to the center. With care, we can save twelve pieces, giving the 38-piece dissection in Figure 15.14. I cut and fold two of the small crosses identically and do the same for another pair. The saving in pieces comes at the loss of some cyclic hinging, as you can verify. I give a perspective view of the five assemblages as they begin to come together in Figure 15.15.

Harry Langman (1950) defined what he termed the *double-cross*, which consists of two Greek Crosses glued together.[‡] He gave a simple, 4-piece unhingeable dissection of this figure to a square. The dissection is a straight-forward application of the 4-piece dissection of two Greek Crosses to a square by Sam Loyd (*Tit-bits*, 1897b). Loyd's dissection is swing-hingeable, with an assemblage for each Greek Cross. Yet, Langman's dissection is not swing-hingeable, because the pieces cannot be connected by hinges into a single assemblage. Unable to find a suitable swing-hingeable dissection

---

[‡]The double-cross is reminiscent of the "double dagger" symbol used for footnotes.

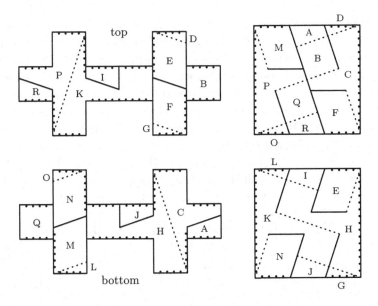

Figure 15.16. Rounded piano-hinged double-cross to a square. (W)

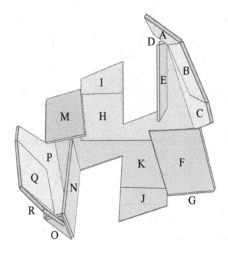

Figure 15.17. Perspective view of a double-cross to a square.

that I could convert to be piano-hinged, I found an 18-piece rounded piano-hinged dissection (in Figure 15.16) by looking for it directly.

It is not difficult to find the side of the square by taking a diagonal of an arm of the double-cross, but it is difficult to see how to fill in the opposite side. Symmetry comes to the rescue if we fold the top level of the neck of the cross (in piece K) down parallel with the bottom level of the neck (in piece H). Pieces C and P fold around to give the square's sides on the top level, yet challenging problems remain. How should we use the material over pieces C and H and under pieces K and P? And how do we fill in the middle of the horizontal sides of the square? The key insight seems to be to cut pieces I and J out of pieces K and H, respectively. This forces us to cut pieces N and E to fill the resulting cavities in pieces K and H. This leaves pieces F and M, which forces us to cut cavities in pieces C and P, respectively. Fortuitously, the pieces from those cavities (pieces A and R) fill out the remaining holes in the horizontal sides of the square.

The dissection is so beautifully symmetrical that you will surely do a double-take the first time you see it. As we see in Figure 15.17, an unexpected bonus is the two pairs of cap-cyclic hingings: (C, D, E, and H), (C, F, G, and H), (K, L, M, and P), and (K, N, O, and P). Yet, if I hadn't been so smug, I might have sniffed out the ultimate double-cross . . . .

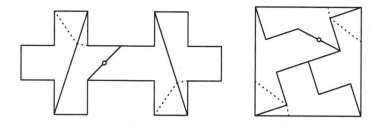

Figure 15.18.   Swing- and twist-hingeable dissection of a double-cross to a square.

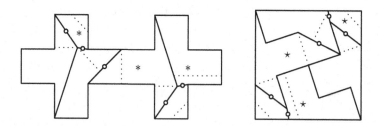

Figure 15.19.   Twist-hinged dissection (plus fold lines): double-cross to square.

Returning to this puzzle after two years, I stumbled upon the 4-piece swing- and twist-hingeable dissection in Figure 15.18. Dotted line segments identify the position of the two swing hinges in the double-cross and in the square. This converts to the 6-piece twist-hinged dissection indicated by the solid lines in Figure 15.19. That dissection then converts to a 13-piece rounded piano-hinged dissection, using the conversion technique in Chapter 4. The dotted lines in Figure 15.19 indicate the positions of piano hinges that flip pieces out of the way. Considering the lovely wooden model for Figure 15.16 that now has too many pieces, it's enough to drive a dissectionist to use some unfortunate double negatives.

Before that last twist of fate, I had gotten ambitious and tried to dissect a Cross of Lorraine to a square. Because the cross consists of thirteen small squares, and $\sqrt{13}$ is the length of a diagonal in a (2×3)-rectangle, chances were good that I would enjoy some success. In fact, a Mr. Szeps and Bernard Lemaire, a professor of operations research in Paris, each found 7-piece unhingeable dissections, as reported by the French math puzzle columnist Pierre Berloquin (*Monde*, 1974a and 1974b). Subsequently, I found a 7-piece swing-hingeable dissection.

We see that the Cross of Lorraine is an $(a^2 + b^2)$-omino for $a = 2$ and $b = 3$. This means that there is a piano-hinged dissection of the cross to a square, using the dissection of any k-omino to any other k-omino or square that we saw in Chapter 11. If we fold the loop to fill in the cross, combining isosceles right triangles along edges not on the boundary of the cross, we find that we use 28 pieces, fourteen on each level. To produce a square, we must fold the loop twelve more times. This gives a total of 40 pieces.

Again, we can do better with a specialized approach. If my swing-hingeable dissection were hinge-snug, we could have converted the swing-hinged dissection to a twist-hinged and then to a piano-hinged dissection. Unfortunately, my swing-hinged dissection is not hinge-snug. So I tried again and found an 11-piece hinge-snug swing- and twist-hingeable dissection. This led to an 18-piece twist-hinged dissection, which led to a 37-piece piano-hinged dissection. That had marginally fewer pieces, but at a significant loss of symmetry. It wasn't until I had had my first success with the double-cross in Figure 15.16 that I saw how to push through the 24-piece piano-hinged dissection in Figure 15.20.

Just as I hinged together pieces H and K in Figure 15.16, I have hinged pieces E and J in Figure 15.20. Similar to the cap-cyclic hinging in Figure 15.16 are the two cap-cycles (pieces J, K, L, and M, and pieces J, N, O, and P) in Figure 15.20. Corresponding to pieces such as A, I, J, and

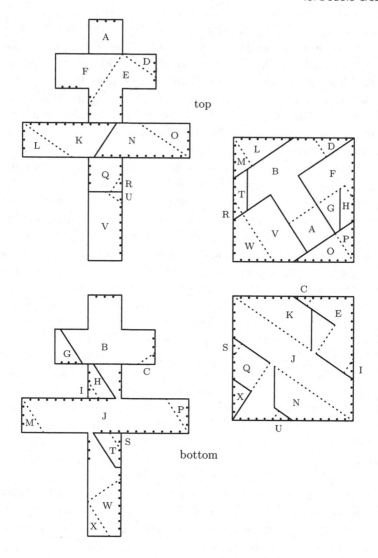

Figure 15.20. Rounded piano-hinged Cross of Lorraine to a square.

R in Figure 15.16 is piece G and pairs of pieces H, I and S, T in Figure 15.20. As is evident in Figure 15.21, the smaller degree of symmetry in
the Cross of Lorraine makes the job tougher and increases the number of
pieces. There's no double standard here for the French, merely acceptance
of an asymmetrical reality. Yet, as just compensation, we get four different
flat-cycles. Can you spot them?

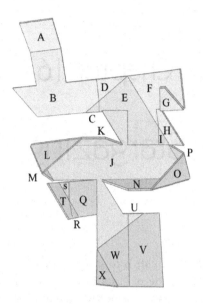

Figure 15.21.   Assemblage for a Cross of Lorraine to a square.

At this point, perhaps even the most loyal reader has lost her patience with all of this double-talk—and nary a double entendre for comic relief! Let's shift gears (no double-clutching, mind you) as we move into the next chapter. We shall focus on those most heavenly of objects, the stars. And I promise that there will be no more double-dealing, though there may be a binary system or two.

Chapter 16

Stargazing

A transcendent experience in life is to stand outdoors on a crisp, clear night, looking up at that multitude of stars in the heavens. Our brief moment in time and our tiny outpost in space are dwarfed as we marvel at those enormous, lustrous objects, spread across the vastness of the universe and into the far distant past. Geometry has its wonders too, spread across the vastness of our imaginations. Such beautiful stars! Are they more eternal than their namesakes?

Like the regular polygons, the geometric stars have a lovely structure based on rhombuses. Harry Lindgren noted that certain families of pairs of stars have simple trigonometric relationships between their dimensions, making possible dissections with relatively few pieces. Let's acknowledge the obvious and call these dissections *auspicious*. In the simplest of cases, such as the first dissection in this book, we may also see a polygon or two sneak in.

If our nascent astronomers now train their naked eyes towards the stars, they may first identify polygons—er, planets—rather than stars. Indeed, they have already seen an auspicious dissection of two triangles to a hexagon in Figure 1.13. This should lead them to also expect a dissection of two hexagons to a triangle. Freese (1957b) gave a symmetric 6-piece unhingeable dissection for this case. I have identified a 10-piece nettable dissection (Figure 16.1). Not surprisingly, the dissection derives from overlaying a tessellation of triangles with a tessellation of hexagons. Aside from the hinge that connects pieces A and B, the triangle possesses a lovely symmetry, as we can see in the perspective view in Figure 16.2. There is another way to hinge pieces E through J. Do you see it?

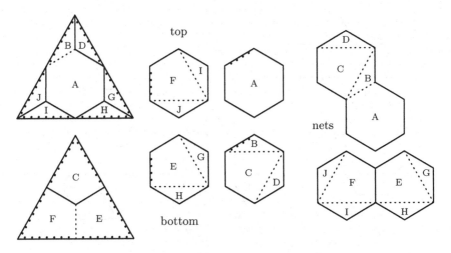

top

nets

bottom

Figure 16.1. Folding dissection of two hexagons to a triangle.

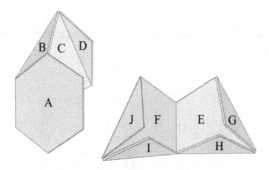

Figure 16.2. Perspective view of assemblages for two hexagons to a triangle.

With galactic impatience, our budding astronomers clamor for their turn at the telescope, so that they might observe something truly stellar. There should be an auspicious dissection of two hexagons to a hexagram. Lindgren (1964b) gave a symmetric 4-piece unhingeable dissection for this case. I have found an 8-piece nettable dissection (Figure 16.3) in which I cut and fold both hexagons identically. Note that each hexagon is tube-cyclically hinged. The symmetry and cyclic hingeability stand out in Figure 16.4.

Perhaps we should polish up our lenses, for we seem to be observing a double image: There is also an auspicious dissection of two hexagrams to

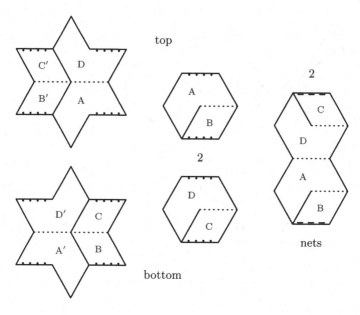

Figure 16.3. Folding dissection of two hexagons to a hexagram. **(C)**

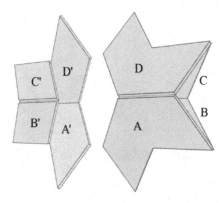

Figure 16.4. Perspective view of assemblages for two hexagons to a hexagram.

a hexagon, for which Lindgren (1964b) gave a symmetric 6-piece unhinge-able dissection. I have found a 10-piece rounded piano-hinged dissection (Figure 16.5) in which I cut and fold both hexagrams in the same way. The key is to have two pieces (C and E) that each have a 60°-rhombus on a

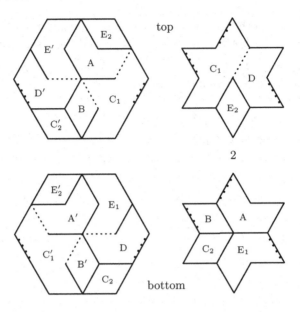

Figure 16.5. Rounded piano-hinged two hexagrams to a hexagon. (C)

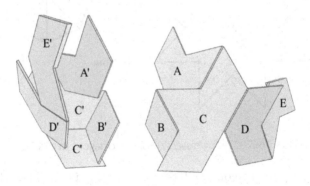

Figure 16.6. View of assemblages for two hexagrams to a hexagon.

second level, so that piece E fills in a corresponding cavity in the hexagram, and piece C does so in the hexagon. The hexagon has an interesting "top versus bottom" symmetry, although it is hard to see this in Figure 16.6.

If we want a dissection in which each piece is only on one level, then it appears that we need 12 pieces. However, we can find such a dissection in which we cut both hexagrams in the same way, and the dissected and folded hexagon has both rotational symmetry and the inside-out property.

Puzzle 16.1. *Find a 12-piece folding dissection of two hexagrams to a hexagon in which each piece is only on one level.*

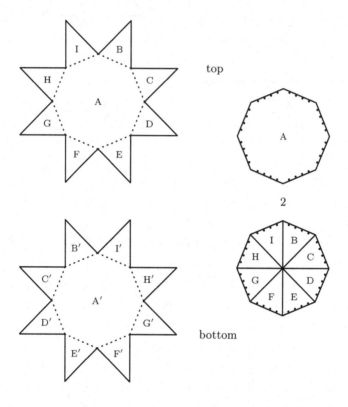

top

2

bottom

Figure 16.7. Folding dissection of two octagons to an {8/3}.

Our next phenomenon exhibits a bit of solar flare: The relationship between hexagons and hexagrams is an example of a class of relationships, identified by Lindgren (1964b), between star $\{(2n + 2)/n\}$ and the corresponding polygon $\{2n+2\}$, for any positive integer n. For the case of $n = 3$, Harry Langman (1962) posed the puzzle of finding a 9-piece dissection of an $\{8/3\}$ to two octagons. Remarkably, this dissection appeared as part of a larger construction in the anonymous Persian manuscript, *Interlocks of Similar or Complementary Figures*, from around 1300.

The corresponding piano-hinged version of this dissection takes eighteen pieces, as I show in Figure 16.7. We cut and hinge each of the two octagons

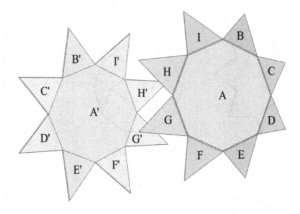

Figure 16.8. Perspective view of assemblages for two octagons to an {8/3}.

in the same way, so that each folds out to form one level in the {8/3}, as we
see in Figure 16.8. This same approach works in general to give a $(4n+6)$-
piece nettable dissection of a $\{(2n+2)/n\}$ to two $\{2n+2\}$s. Readers may
enjoy sketching the case for $n = 4$, namely, of a {10/4} to two decagons.
Interestingly, there is more than one way to hinge these dissections.

Puzzle 16.2. *Find another way to piano-hinge the dissection of an {8/3} to
two octagons.*

The next wonder seems to have coalesced from interstellar dust. Lindgren (1964b) gave a 6-piece unhingeable dissection of a {10/3} to two
decagons. His dissection relies on the internal structure of the {10/3} and
the decagon and cuts these figures only on the boundaries of the constituent
rhombuses. The 14-piece folding dissection in Figure 16.9 similarly relies
on the internal structure but is a bit more complicated. In particular, I
force the large pieces (A and B) on either level of the {10/3} to be adjacent
so that I can hinge them together. We can then fill in a rhombus on the
exterior of the {10/3} with a rhombus (piece F) from the interior of the
decagon. As we see in the figure, the dissection is nettable. Do the assemblages in Figure 16.10 remind you of butterflies? (Surely not butterflies in
the zodiac!)

This dissection is less of a cosmic anomaly than we might at first think.
In fact, it's an instance of a general relationship, of two $\{4n+2\}$s to a
$\{(4n+2)/(n+1)\}$, where $n = 2$. For the next smaller value, $n = 1$, the relationship is of two hexagons and a hexagram, which we saw in Figure 16.3.

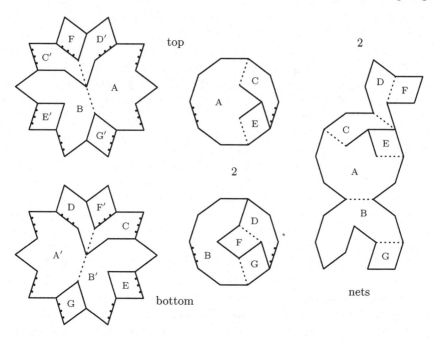

Figure 16.9. Rounded folding dissection of two {10}s to a {10/3}. (C)

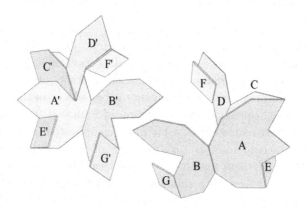

Figure 16.10. Perspective view of assemblages for two {10}s to a {10/3}.

For the next larger value, $n = 3$, the relationship is between two {14}s and a {14/4}. Figure 16.9 hints at some strategies for solving this puzzle,

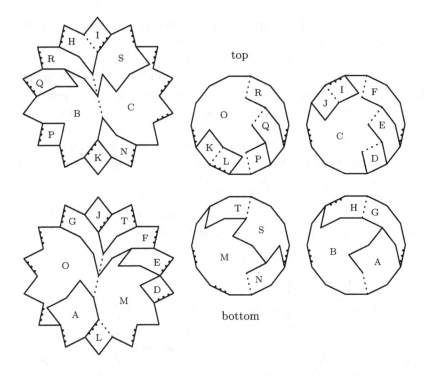

Figure 16.11. Rounded folding dissection of two {14}s to a {14/4}.

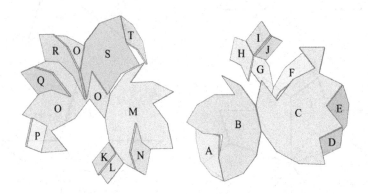

Figure 16.12. Perspective view of assemblages for two {14}s to a {14/4}.

but experimentation quickly indicates that it is not so easy to apply those strategies.

After a good deal of trial and error, I devised the 20-piece rounded piano-hinged dissection in Figure 16.11. Cut the two {14}s similarly, but not identically. There are four main pieces, with pieces C and O similar and pieces B and M similar. Pieces P, Q, R, and T are the same as pieces D, E, F, and H, respectively. Yet, to create pieces I and J to position at the upper point of the {14/4}, and pieces K and L at the lower point, we need to break the symmetry. Note that pieces N and S are similar to pieces A and G, but the cavity in O created by removing K and L forced me to fatten piece A, whereas the cavity in piece C forced me to fatten piece S. Do the new assemblages in Figure 16.12 suggest to you mutated butterflies? (What would we do to the zodiac?!) For the first three cases of the general relationship, $n = 1, 2, 3$, we have used $6n+2$ pieces. Are $(6n+2)$-piece folding dissections possible for all values of n?

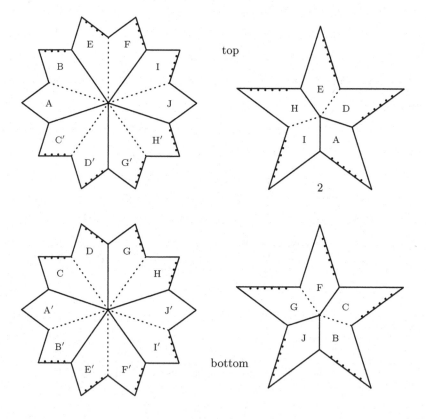

Figure 16.13.   Folding dissection of two {5/2}s to a {10/3}. (W)

The next astronomical event is as brilliant as the birth of a supernova. Ernest Freese (1957a) identified a 10-piece dissection of a $\{10/3\}$ to two pentagrams. Lindgren (1964b) independently discovered this dissection and furthermore identified a special trigonometric relationship, which I listed in my first book, between a $\{(4n+2)/(2n-1)\}$ and a $\{(2n+1)/n\}$, for any positive integer n. For $n = 2$, it describes the dissection of a $\{10/3\}$ to two pentagrams. This relationship leads to a $(4n+2)$-piece dissection of a $\{(4n+2)/(2n-1)\}$ to two $\{(2n+1)/n\}$s. These lovely dissections are swing-hingeable. However, the real surprise is that their two-level-thick generalizations are also fold-hingeable, yielding an $(8n+4)$-piece dissection.

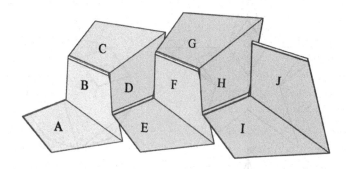

Figure 16.14.   Perspective view of one assemblage for two $\{5/2\}$s to a $\{10/3\}$.

We can see this for the case of two pentagrams to a $\{10/3\}$ in Figure 16.13. We hinge the ten identical pieces from a pentagram alternately on a side of the pentagram and then a side of the $\{10/3\}$. As we see in Figure 16.14, the hinging gives the model an extraordinary pleated appearance as it forms an apparent "wall of stars." This dissection also has the inside-out property. If we use two colors for the pieces, then there is a lovely color arrangement. If we take a different color for each level of the pentagrams, then the pieces alternate colors on each level of the $\{10/3\}$. Yet another nifty feature is that the ten pieces in each pentagram fold up to form a stack, going from piece A to piece J in alphabetical order.

Freese also identified the related 8-piece swing-hingeable dissection of an $\{8/2\}$ to two $\{4/1.5\}$s and 12-piece swing-hingeable dissection of a $\{12/4\}$ to two $\{6/2.5\}$s. We call $\{4/1.5\}$ and $\{6/2.5\}$ pseudostars. A *pseudostar* $\{p/q\}$, where $1 < q < p/2$ and q is not a natural number, is a starlike figure in which the boundaries out of a vertex do not head directly towards another vertex. Imagine a circle with p equally-spaced vertices on it. The boundaries from a vertex head toward positions on the circle that are q/p of the way around the circle, in a clockwise and in a counterclockwise direction.

Freese's dissections generalize, along with the previous relationship, to lead to a $(2n+4)$-piece swing-hingeable dissection of two $\{(n+2)/((n+1)/2)\}$s to a $\{(2n+4)/n\}$. The corresponding fold-hingeable dissection has $(4n+8)$ pieces. When $n = 2$ we get the dissections of two $\{4/1.5\}$s to an $\{8/2\}$, and when $n = 4$ we get the dissections of two $\{6/2.5\}$s to a $\{12/4\}$. Are these dissections of pseudostars the geometric equivalent of exotic nebulas?

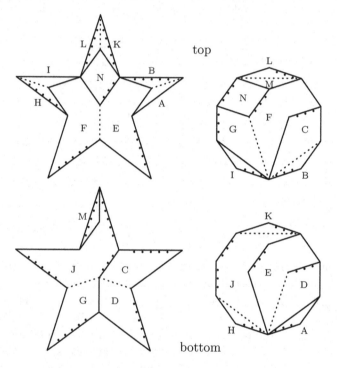

Figure 16.15.   Hyper-extended piano-hinged pentagram to a decagon.

Lindgren (1964b) gave a lovely 6-piece unhingeable dissection of a pentagram to a decagon. He packed three of the points of the pentagram into the decagon in a fashion somewhat similar to how he packed the points of the pentagram into the $\{10/3\}$. Thus, we can draw inspiration from the way we have piano-hinged the pieces in Figure 16.13 to produce the 14-piece folding dissection in Figure 16.15. Too bad that its inside-out property is hard to see in Figure 16.16. Also, the hinge that connects pieces J and N is

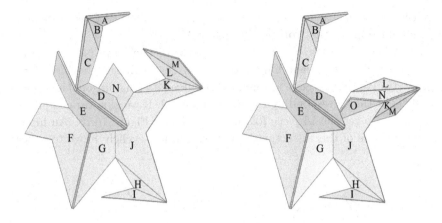

Figure 16.16. Hyper-ext. {5/2} to {10}.   Figure 16.17. Folding {5/2} to {10}.

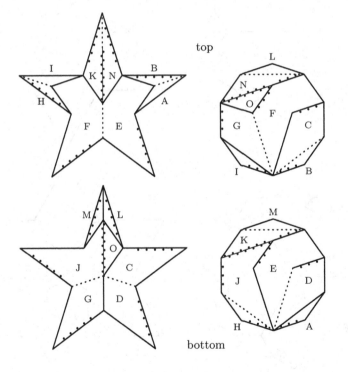

Figure 16.18. Folding dissection of a pentagram to a decagon. (W)

nonstandard, in that we use it to rotate piece J by 360°. A *hyper-extended* piano hinge is one that rotates through 360°.

But here's the big bang: At the expense of one more piece, we can produce a properly piano-hinged dissection that still has the inside-out property. We split piece N in Figure 16.15 along its long diagonal and similarly split the corresponding rhombus in piece J. If we glue various pieces together, we get the 15-piece folding dissection in Figure 16.18. An interesting feature of this dissection is that we can leaf-cyclicly hinge together

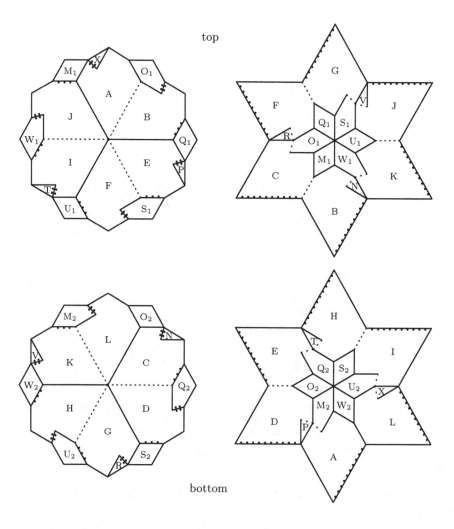

Figure 16.19. Rounded piano-hinged dissection of a {12/2} to a {6/2}.

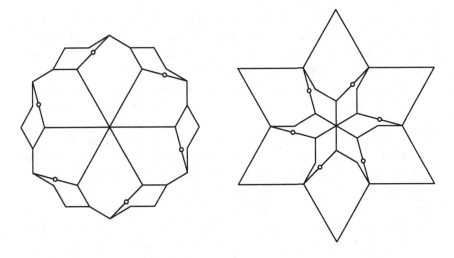

Figure 16.20.   Swing- and twist-hingeable dissection of a {12/2} to a {6/2}.

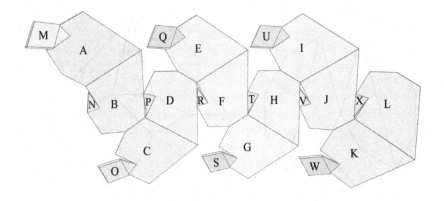

Figure 16.21.   Perspective view of a {12/2} to a {6/2}.

pieces N, O, J, and K. With high-power magnification, we can detect that cyclic hinging in Figure 16.17.

Lindgren (1964b) identified a special trigonometric relationship, which I listed in my first book, between a $\{(4n+4)/2\}$ and a $\{(2n+2)/n\}$, for any positive integer n. For $n = 2$, it leads to a 9-piece unhingeable dissection of a {12/2} to a {6/2}. The same basic approach gives a 12-piece swing-hingeable dissection in my second book (2002), which we can turn into the

12-piece swing- and twist-hingeable dissection in Figure 16.20. Since I do not convert the swing hinges here, I have not identified them in the figure. Let's take this as our starting point for a piano-hinged dissection. Note that we use half of a 30°-rhombus to assist with the twist, and we twist at the midpoint of the rhombus' short diagonal, along which we have cut.

We can use the same basic approach as in Figure 16.13 to take the points of the hexagram and pack them into the interior of the {12/2}. We piano-hinge the 12 large pieces (A–L) from the hexagram alternately on a side of the hexagram and on an interior side. Then, we convert the twist hinges in Figure 16.20 using a variation of our conversion technique in Chapter 4. The small, narrow isosceles triangles are each one half of the rhombus that we mentioned in Figure 16.20. They piano-hinge along the short diagonal of the rhombus so that, for example, piece X hinges with piece L, piece P with piece D, and so on. The final result is the 24-piece rounded piano-hinged dissection in Figure 16.19. This model also has an extraordinary pleated appearance, as we see in Figure 16.21. Is this our "great wall" of stars?

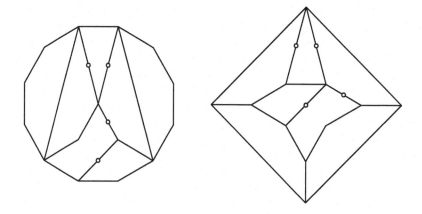

Figure 16.22. Swing- and twist-hingeable dissection of a dodecagon to a square.

Starry-eyed readers may be a bit disappointed with our final dissection here, as we focus in on a dodecagon and a square, neither of which is a star. Yet, this is not really a disaster, since we will use techniques common to this chapter. Lindgren (1951) discovered a lovely 6-piece unhingeable dissection of a dodecagon to a square. I adapted it into a 9-piece twist-hingeable dissection in my second book (2002). Of course, we could convert it into an 18-piece rounded piano-hinged dissection, using the conversion

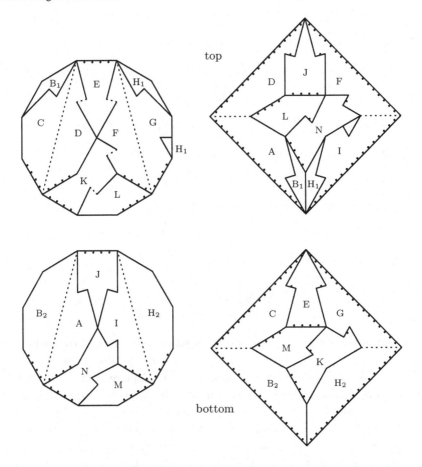

Figure 16.23. Rounded piano-hinged dissection of a dodecagon to a square.

technique in Chapter 4. However, we can do a bit better if we use the 7-piece swing- and twist-hingeable dissection in Figure 16.22. We will use the conversion technique on the four twist hinges but use a different technique to connect the rest together. Since I do not convert the swing hinges, I have once again not identified them in the figure.

I take advantage of the near-symmetry of the four long pieces in Figure 16.22, trying to use them as though they were interchangeable. In Figure 16.23 I piano-hinge such pieces that are adjacent in the dodecagon so that they are stacked on top of each other in the square. This approach requires piece C to be the same shape as piece A and also piece G to be the same shape as piece I. To do this I then need piece B to absorb what I have

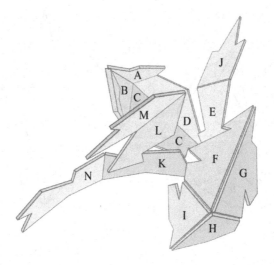

Figure 16.24.   Perspective view of a dodecagon to a square.

removed from piece C and also piece H to absorb what I have removed from piece G. The result is a 14-piece rounded piano-hinged dissection. Because it is so small, I do not label the triangular part of piece H in the top level of the square. Judging from the cardboard model that I constructed, it seems that this dissection still works if we employ two cap-cycles: (A, B, C, D) and (F, G, H, I). We marvel at this last wonder in Figure 16.24.

Yet, at this point, with no more stars on the horizon, we seem to have exhausted our solar furnaces. Even so, let's hope that this quick survey of our geometric universe has put a twinkle in your eyes and a solar wind in your sails.

Chapter 17

Manifold Blessings

How lovely when we can take multiple copies of some figure, fold them this way and that, then assemble them into a larger version of the same figure. Besides the hexagons, triangles, and square of Chapter 10, we will now handle Greek Crosses, Latin Crosses, hexagrams, and dodecagons. In taking stock of these additional piano-hinged dissections, let us count our blessings. Every one of these intriguing dissections, except for the last, involves at least two cyclic hingings, and altogether they exhibit four different types of cyclic hinging: tube-cyclic, leaf-cyclic, cap-cyclic, and saddle-cyclic. And not to disappoint, the last dissection will display terrific symmetry.

The easiest of these dissections are of four to one. Since a Greek Cross consists of five squares, what could be easier? Ernest Freese (1957b) displayed an unhingeable dissection that has just eight pieces. To produce a 16-piece piano-hinged dissection, we cut and hinge each small cross identically, as in Figure 17.1. Further evidence of our good luck is that each small cross is tube-cyclicly hinged, as we see in Figure 17.2.

Tougher is the Latin Cross, which consists of six squares. While an unhingeable dissection needs no more than eight pieces, we produce the 17-piece piano-hinged dissection of Figure 17.4. We cut and hinge one of the small Latin Crosses (pieces A, B, C, and D) in a fashion analogous to a small Greek Cross in Figure 17.1. We cut another (pieces E, F, G, and H) in a mirror-image fashion, and we use the two to fill in the head and most of the two arms of the large cross. A different variation of our cutting and hinging of a small Greek Cross fills in the middle of the large cross (pieces I, J, K, and L). With a bit more work and five more pieces, we can cut the fourth small cross to fill in the rest of the trunk of the large cross. See

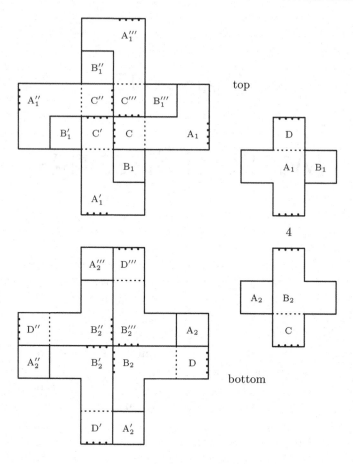

Figure 17.1. Piano-hinged dissection of four Greek Crosses to one.

how blessed we are in Figure 17.3 with tube-cyclicly hingings in three of the small crosses.

Easier again is the hexagon. An unhingeable dissection of four hexagons to one needs just six pieces, a swing-hingeable dissection needs at most seven pieces, and a twist-hinged dissection needs at most eight pieces. There is a simple 10-piece piano-hinged dissection (Figure 17.5). We leave two of the hexagons uncut and cut two others into four pieces each. What luck that these two sets of pieces are tube-cyclicly hinged, as we see in Figure 17.6!

Since hexagrams consist of triangles, we know to expect an advantageous dissection of three to one. Ernest Freese (1957b) and Harry Lindgren

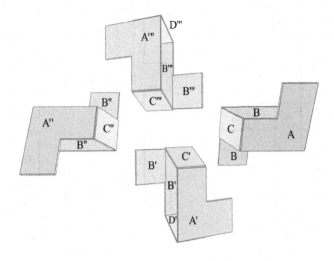

Figure 17.2. View of hinged pieces for four Greek Crosses to one.

(1964a) independently found a lovely 12-piece dissection, which turns out
to be swing-hingeable. Both Freese and Lindgren cut all three small stars
identically, with all cuts having one endpoint in the center. What a surprise for me to discover that this dissection can be used on both levels of

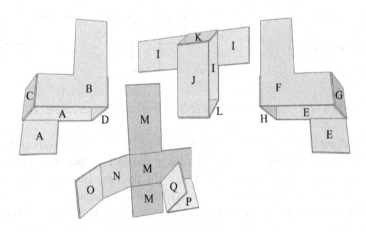

Figure 17.3. View of hinged pieces for four Latin Crosses to one.

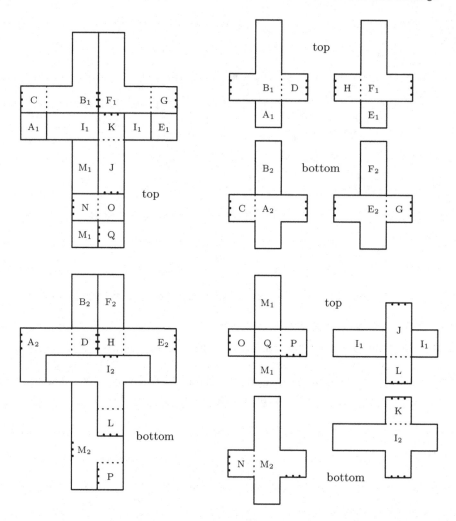

Figure 17.4.   Piano-hinged dissection of four Latin Crosses to one.   (C)

a rounded piano-hinged dissection (Figure 17.7)! How fortuitous to have a cap-cyclic hinging that involves both large pieces together with two small triangles! To convert the small star into the configuration that fits in the large star, first fold out pieces B and G, then manipulate pieces C, D, E, and F of the cap, and finally fold down pieces A and H. We easily slide the three resulting configurations together to give the large star. In Figure 17.8 I have shown the three identical assemblages from three different vantage points. The middle one is looking from the top down.

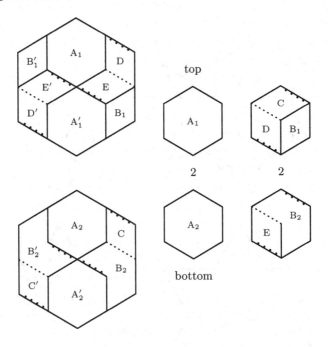

Figure 17.5.   Piano-hinged dissection of four hexagons to one.

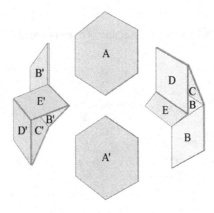

Figure 17.6.   View of hinged pieces for four hexagons to one.

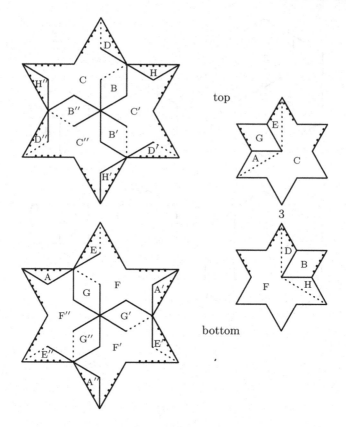

Figure 17.7.   Piano-hinged dissection of three hexagrams to one. (W)

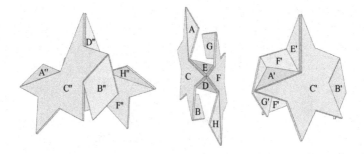

Figure 17.8.   View of three hexagrams to one.

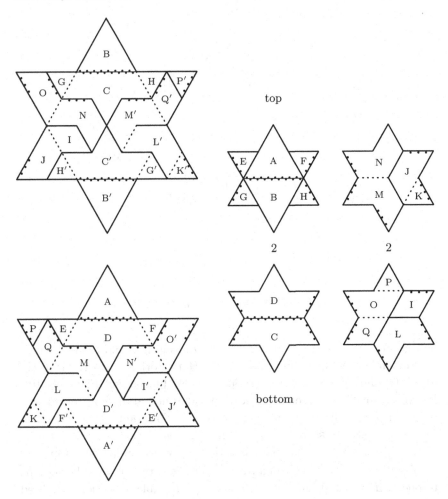

Figure 17.9. Folding dissection of four hexagrams to one. (C)

We next consider a dissection of four hexagrams to one. My 34-piece folded dissection is in Figure 17.9. I cut and hinge a pair of small hexagrams identically and cut and hinge the other pair identically. The challenge in this dissection is to get some of the points of the small hexagrams out of the way when forming the large hexagram. To accomplish this, I use a leaf-cyclic hinging on the pieces A, B, C, and D. Then, the pieces E, F, G, and H move from being next to pieces A or B to being next to pieces C or D. Thus, we can fill in symmetric sections at the top and bottom of the large

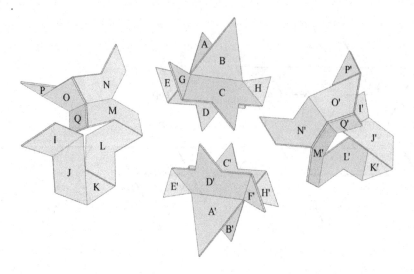

Figure 17.10.   View of four hexagrams to one.

hexagram. This leaves symmetric sections on the right and the left, which by trial and error I figured out how to fill. Hard work is rewarded as pieces M, N, O, and Q turn out to be saddle-cyclicly hinged (see Figure 17.10). I have arranged the assemblage so that the dihedral angles between the pieces in the saddle-cyclic hinging are equal. The common angle is then $180° - \arccos(1/3) \approx 109.4712°$. With the leaf-cyclic and saddle-cyclic hingings, it is no fluke that this dissection enjoys the inside-out property.

We also know to expect an advantageous dissection of seven hexagrams to one. I have based mine on a 19-piece unhingeable dissection by Alfred Varsady. I fold out one of the small hexagrams to fill a hole in the bottom level of the large hexagram. Then, I cut and fold three of the remaining small hexagrams symmetrically and arrange them symmetrically in the large hexagram. These constitute pieces E, F, G, H, I, and J, which I hinge with a saddle-cycle among pieces F, G, H, and I. Finally, I cut and fold the three remaining small hexagrams identically and arrange them symmetrically in the large hexagram.

This last step is a bit trickier, since it calls for filling in the center of the top level of the large hexagram, which I do with pieces K and L. Piece L steals area from the top of the small hexagram that I would have liked to use for pieces N and P. To compensate, I fold over a 60°-rhombus from the bottom level of the small hexagram and continue the cut/fold line

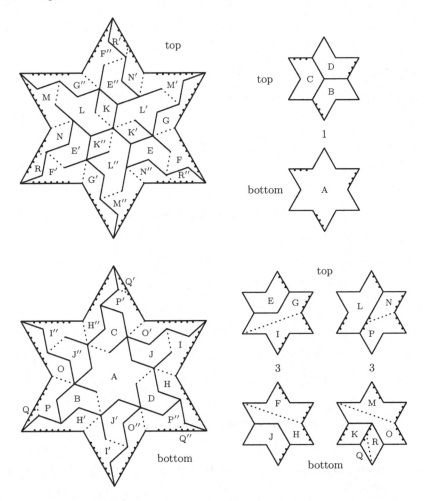

Figure 17.11. Folding dissection of seven hexagrams to one. (C)

between pieces N and P to produce pieces Q and R. Our boon from these manipulations is that the resulting three assemblages have both a saddle cycle (pieces M, N, O, and P) and a flat cycle (pieces N, P, Q, and R).

My 46-piece folded dissection in Figure 17.11 has a lot of pieces. Yet, converting a 25-piece swing-hinged dissection would bring even more. In Figure 17.12 I show an assemblage for each type of small hexagram, with each positioned roughly as the labeled pieces appear in Figure 17.11. Again, I have chosen the dihedral angles between the saddle-cyclicly hinged pieces

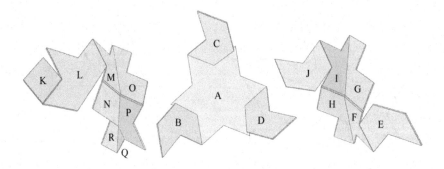

Figure 17.12.   View of (three of the) seven hexagrams to one.

Figure 17.13.   Three of the seven hexagrams to one (with middle flippant).

to be equal, which forces them to be $180° - \arccos((16 - 5\sqrt{7})/9) \approx 107.93368°$.

We can do a bit better (Figures 17.13 and 17.14) if we use a certain nonstandard hinge. A *flippant* piano hinge is one that attaches an edge of one piece to a surface of another piece, allowing a rotation of 180°, but keeping the rotated piece on the same level. As in Varsady's dissection, we cut six of the seven small stars identically. We cut each level into three pieces, in such a way that we can hinge pieces F, G, H, and I in a saddle-cyclic manner. We fit the six small stars around the exterior of the large star, leaving a hole in the shape of a hexagram on each level.

Although these holes have a common center, they are rotationally offset from each other. However, we can split one level of the seventh star into four pieces, and then fold them over. Fold pieces B and C to be perpendicular to piece K, and similarly with pieces D and A. Then, fold C flat against B,

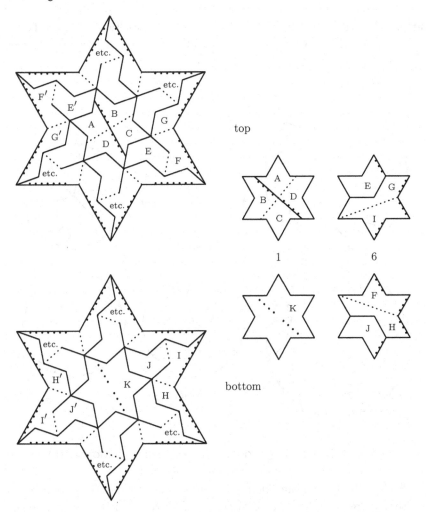

Figure 17.14. Flippant and standard piano-hinged dissection of seven hexagrams.

and A flat against D. Fold piece B flat against K and 180° from where it started, and similarly handle piece D. Fold piece C against K, and similarly with A, and we are done. Of course, we are using a flippant piano hinge to connect pieces B and D to piece K. In return, we get a dissection with only 41 pieces, which also contains six lovely saddle-cyclic foldings.

You might question whether it is fair to use flippant hinges. I am not sure that it is, even though thin paper does seem to allow us to play with such hinges, if we want. However, I did not discover the dissection in Figure 17.11 until a year after I had discovered the dissection in Figure 17.14, and I doubt that I would have the good fortune to discover the former without having first figured out the latter.

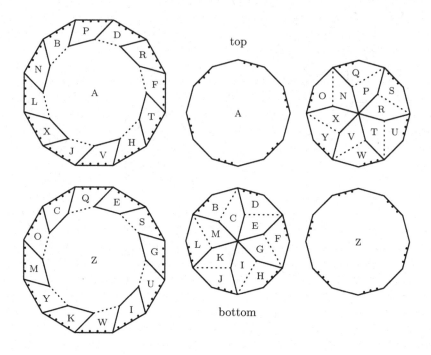

Figure 17.15. Piano-hinged dissection of two dodecagons to one. (C)

So far, all of the geometric figures in this chapter have been built from either squares, in the case of Greek Crosses or Latin Crosses, or equilateral triangles, in the case of hexagons or hexagrams. Yet, we are not restricted to only these figures, as we shall see with our final dissection of the chapter, the 26-piece piano-hinged dissection of two dodecagons to one in Figure 17.15. The dissection relies on a 13-piece unhingeable dissection by Joseph Rosenbaum (1947), with each level in the large dodecagon being a copy of the large dodecagon in Rosenbaum's dissection. Note that each of the 12 identical pieces on the appropriate levels of each small dodecagon consists of a small equilateral triangle and half of a small square.

Since Rosenbaum's dissection has sixfold rotational symmetry, it is not surprising that this piano-hinged dissection does too. How serendipitous that a perceived imperfection in Rosenbaum's dissection, namely the neces-

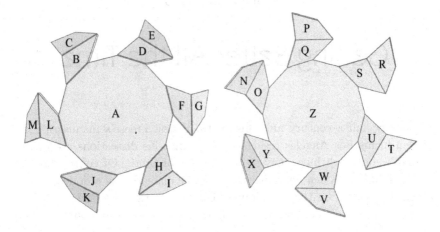

Figure 17.16.   View of piano-hinged two dodecagons to one.

sity of turning over six of the pieces, is the feature that makes our dissection work. I have not seen how to generalize the 11-piece unhingeable dissection of Freese (1957b) or the 10-piece unhingeable dissection of Lindgren (1964b) into a corresponding piano-hinged dissection. The wonderful simplicity of our dissection, which we enjoy once again in Figure 17.16, is the perfect note upon which to conclude so successfully this chapter.

Manuscript 5

Getting Better All the Time

For nearly half a century after his death, Ernest Freese's manuscript lay hidden in his Los Angeles residence. Now that his dissections are finally available, how well have they stood the test of time? Of course, no one has been able to touch his remarkable 6-piece dissection of a dodecagon to a hexagon, based on the technique of completing the tessellation. Of the nine of Freese's dissections that Lindgren reproduced in his book, people have improved four others over the years. Probably their appearance in Lindgren's book played a key role in the improvement of two of those four. These are his five-into-one dissection of pentagrams and his 9-piece dissection of an enneagon to a triangle. These improvements, the first by me and the second by Gavin Theobald, appear in my first book (1997).

Of the remaining approximately 320 dissections, approximately two dozen have seen improvements in the 45 years during which Freese's manuscript was unavailable. On some, Freese just barely got beaten, such as his 11-piece dissection of two dodecagons to one. Freese had a great idea on that one, but didn't push it all the way through, as Harry Lindgren (1964b) did with his 10-piece dissection. On others, Freese was soundly beaten, although of course no one knew it when the better dissection appeared. These include

- a $\{12/2\}$ to a hexagram, from 13 to 9 pieces (Lindgren, 1964),

- a $\{12/2\}$ to a hexagon, from 15 to 10 (Lindgren, 1964) to 8 (me, 1972),

- three octagons to one, from 24 to 10 (me, 1974),

- five octagons to one, from 24 to 17 (me, 1997),

- three enneagons to one, from 37 to 21 (Cundy and Langford, 1960), to 18 (Lindgren, 1964) to 15 (me, 1972) to 14 (Hanegraaf and Reid, in my 1997 book),

- three dodecagons to one, from 24 to 15 (Lindgren, 1964) to 14 (Hanegraaf and Reid, in my 1997 book),

- dodecagons of area ratio 1 to 2 to 3, from 25 to 12 (Valens, 1964),

- an octagon to an {8/2} plus an {8/3}, from 9 to 6 (Reid, unpublished, before 1997),

- twelve dodecagons to one, from 78 to 47 (Reid, unpublished, before 1997).

The first six of these are in my first book. The last two, by Robert Reid, are unpublished.

Not only did people have little knowledge of Freese's dissections, but they also had little knowledge of the kinds of dissection puzzles on which he had worked. His manuscript contains solutions to some puzzles that no one had worked on but him. How good are the dissections for these puzzles that are uniquely Freese's?

Some of his dissections are inventive solutions that are unlikely to be improved upon, while others have cried out for friendly competition from another dissection expert. In response, I rolled up my sleeves and, with the benefit of fifty years of progress, found a number of improvements. Altogether, I have found approximately three dozen dissections that can be improved. As a sampling, I provide four improved dissections here. Too bad that Freese isn't around to respond with further improvements!

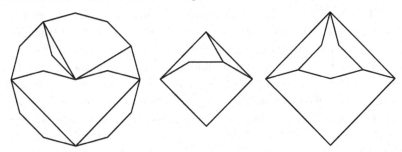

Figure M5.1.   Improved dodecagon to 1-square and √2-square.

In Plate 124 of his manuscript, Ernest Freese gave two different ways to dissect a dodecagon into two squares with an area-ratio of 1 : 2. Both use nine pieces and are relatively symmetric. However, I found a 7-piece dissection (Figure M5.1) that is also reasonably symmetric. It is related to the second of Freese's two dissections and relies on cutting parts off of certain pieces and gluing them onto other pieces. This dissection takes advantage of the internal structure of the dodecagon. Can you sketch the rhombuses and half-rhombuses that lurk within the solution?

I had not seen this particular dissection puzzle before and was inspired to find a swing-hinged dissection that would lead to a twist-hinged dissection. The swing-hinged dissection that I ask for below leads to a 15-piece twist-hinged dissection.

Puzzle M5.1. *Find a 10-piece hinge-snug swing-hinged dissection of a dodecagon to a 1-square and a $\sqrt{2}$-square.*

Figure M5.2.   Improved triangles for $1^2 + 3^2 + 5^2 + 7^2 = (\sqrt{84})^2$.

In Plate 21 (Figure M2.2), Freese gave a 9-piece dissection of four equilateral triangles with side ratios of $1 : 3 : 5 : 7$ to an equilateral triangle of side $\sqrt{84}$. He was justifiably pleased to find that the triangles fit so well together, with the 3-triangle and 7-triangle sitting side by side so that their opposing vertices are exactly $\sqrt{84}$ apart and with a slice of the 5-triangle being half this length. It took Freese just two more cuts on the 5-triangle to get pieces that form the large triangle. However, we need just one more piece, if we consider cutting a larger piece out of the 5-triangle, so that the cavity created accommodates the apex, once we rotate the piece by $60°$. The resulting 8-piece dissection is in Figure M5.2.

In Plate 176 of his manuscript, Freese introduced a "concave hexadecagon" and gave a 13-piece dissection of it to a square. A concave hexadecagon is a 16-sided polygon in which four sequences of four edges bow in rather than bow out, giving a starlike figure. You can tessellate the plane with the pair consisting of a regular hexadecagon and a concave hexadecagon. Freese found that if you clip off the outermost of the four points of the concave hexadecagon and slice each into two right triangles, then the right triangles fill the remaining cavities to produce an octagon with four long sides and four short sides. He then used the completing-the-

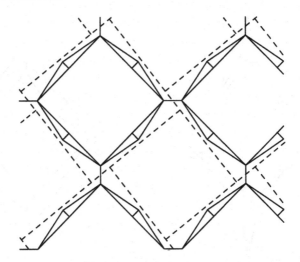

Figure M5.3.   Tessellations for Freese's concave $\{16\}$ to a square.

tessellation technique to convert the octagon to a square, as illustrated in Figure M5.3.

We can avoid slicing the four points into right triangles if we cut two cavities in the square for them to fit into. The pieces that we cut out, two isosceles triangles, help to fill in the open space within the square. I show the resulting 11-piece dissection in Figure M5.4.

In Plate 41 of his manuscript, Freese gave a 9-piece dissection of squares for $13^2 + 19^2 + 25^2 + 31^2 = 46^2$. This was one example of his fascination with number identities in which the numbers on one side of the equation are in an arithmetic progression. We can improve his dissection by one piece, as we see in Figure M5.5. Surprisingly, many features of his dissection remain intact, namely, not cutting the 13-, 19-, and 25-squares, positioning them in the same way in the 46-square, cutting almost the same large piece out of the 31-square, and positioning it in the same way. However, my use of the step technique in this dissection is new, so the way that I fill in the area between the large pieces is completely different.

The four improvements here are a representative sample. Surely, if Freese had lived to see later work on dissections, such as the book by Harry Lindgren (1964b) and even my appendix (1972a), he might have discovered some of these improvements himself. Yet, failing that, we now have a wonderful source book of puzzles upon whose solutions we can try

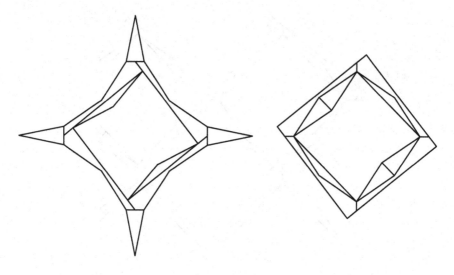

Figure M5.4.   Improved concave {16} to a square.

to improve. Thus, Ernest Irving Freese has left to posterity not only some marvelous dissections but also some stimulating dissection challenges.

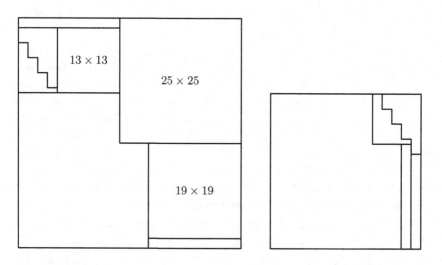

Figure M5.5.   Improved squares for $13^2 + 19^2 + 25^2 + 31^2 = 46^2$.

Chapter 18

A Mixed Bag

Until now, the geometric figures in our dissections have been rather clannish: Except for our chapter on stars, whenever we have dissected two or more figures into another figure, the figures have all been the same shape. Yet there is no good reason why they should stick so closely to their own. We can work to broaden their horizons and show them how to interact with a greater variety of their fellow figures. Then each dissection will be a mixed bag—the manifestation of a mind-expanding, multi-polygonal experience! To keep the number of pieces manageable, let's take advantage of certain natural relationships. In this chapter we will handle triangles, hexagons, and hexagrams.

A good puzzle with which to begin is the dissection of three 1-hexagrams to a 6-triangle. Beneath the surface (or actually, inside the boundaries),

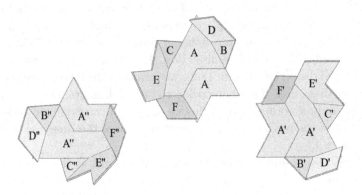

Figure 18.1. Perspective view of three 1-hexagrams to a 6-triangle.

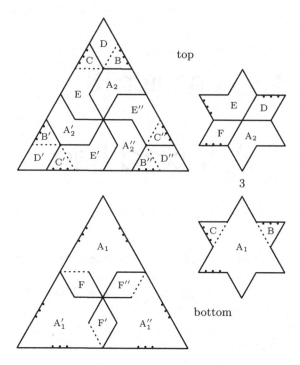

Figure 18.2. Piano-hinged dissection of three 1-hexagrams to a 6-triangle. (C)

we find that these figures consist of the same basic stuff, namely equilateral triangles. Each hexagram contains twelve small triangles, and a 6-triangle contains 36 small triangles. Stuart Elliott (1983) found a 9-piece unhingeable dissection: He cut two points off of each hexagram and packed these points between the remaining pieces to get the triangle. Although Elliott's dissection has lovely symmetry, it is not minimal, as Robert Reid (1987) demonstrated with his 8-piece dissection. However, in finding an 18-piece piano-hinged dissection (Figure 18.2), we return to the threefold symmetry of Elliott's dissection. The main trick is to fit each hexagram into one of the 6-triangle's three corners. To do this we must fold two points of each hexagram out of the way. I show a perspective view of the three assemblages in Figure 18.1.

We can also pose the puzzle the other way around, as three triangles to a hexagram.

Puzzle 18.1. *Find a piano-hinged dissection of three 2-triangles to a 1-hexagram.*

Since we can easily dissect a hexagram to two hexagons, the last dissection points the way towards dissecting six 1-hexagons to a 6-triangle. David Paterson (1989) found a 9-piece unhingeable dissection of six 1-hexagons to a 6-triangle. We can design the piano-hinged dissection in Figure 18.3 by using Paterson's dissection on both levels and piano-hinging appropriately. We could have derived this by superposing tessellations.

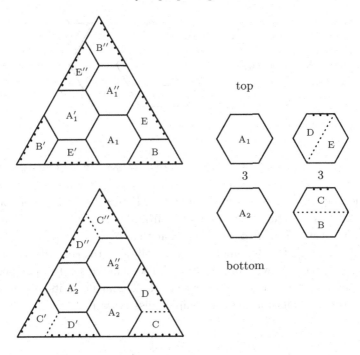

Figure 18.3.   Piano-hinged dissection of six 1-hexagons to a 6-triangle.

Our next dissection integrates both large and small, demonstrating that we are anything but sizist! A 2-hexagon consists of 24 small triangles, and a 5-triangle has 25 small triangles. Thus, adding a 1-triangle to the former yields the latter. There are at least two simple 4-piece dissections of a 1-triangle plus 2-hexagon to a 5-triangle. One of these is swing-hingeable. We can use the other as a basis for a 7-piece piano-hinged dissection, as in Figure 18.4. The pieces from the hexagon come in symmetrical pairs, which fold into the 5-triangle in Figure 18.5.

We can dissect a $\sqrt{3}$-hexagon into eight 1-triangles. Still championing the diminutive, we see that adding a 1-triangle produces the ingredients of a 3-triangle. Using a simple 5-piece swing-hinged dissection as a basis, we

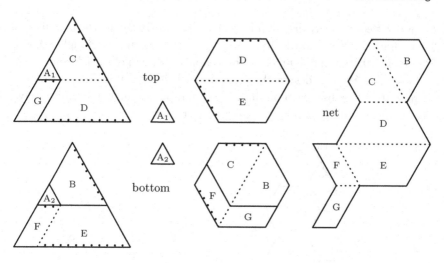

Figure 18.4. Piano-hinged 1-triangle and a 2-hexagon to a 5-triangle. (C)

find the 12-piece piano-hinged dissection in Figure 18.7. The net for the
√3-hexagon is in Figure 18.8. The pieces from the hexagon come primarily
in symmetrical pairs, with the exception that piece K matches up with
pieces J and H. In the 3-triangle, piece L will fit in the concave portion of
piece J, and piece H will fold on top of piece G. We see the perspective
view in Figure 18.6. (This is larger than the 5-triangle to its left in order
to show more detail in this more complex dissection.)

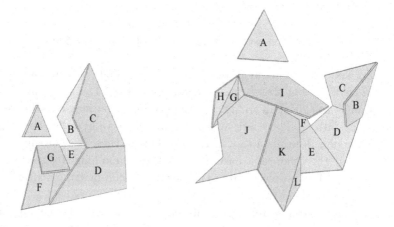

Figure 18.5. Forming a 5-triangle.   Figure 18.6. Forming a 3-triangle.

Although supportive of unusual combinations, we do not wish to (reverse) discriminate against the more standard groupings of triangles and hexagons, either. A 1-hexagon consists of six small triangles, and a $\sqrt{7}$-triangle is equivalent to seven small triangles. Taking advantage of this, Paterson (1989) found a 5-piece dissection of a 1-triangle and a 1-hexagon to a $\sqrt{7}$-triangle. It derives from a superposition of tessellations and turns out to be swing-hingeable. We can adapt it in a simple way to give the 10-piece folding dissection in Figure 18.9. The pieces in the 1-triangle are cap-cyclicly hinged, and the pieces in the 1-hexagon are tube-cyclicly hinged, as we see in Figure 18.10.

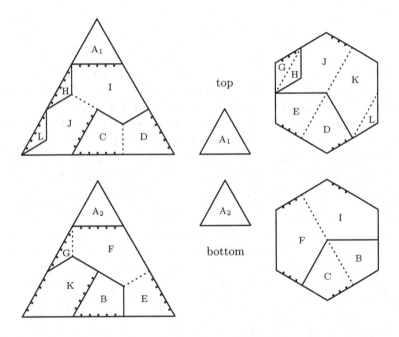

Figure 18.7. Piano-hinged 1-triangle and a $\sqrt{3}$-hexagon to a 3-triangle. (C)

Similarly, we can adapt Paterson's 7-piece dissection of a 1-triangle and two 1-hexagons to a $\sqrt{13}$-triangle. It too derives from a superposition of tessellations but one that is this time not swing-hingeable. In a straightforward fashion we can get the 14-piece folding dissection in Figure 18.12. The pieces in the 1-triangle are cap-cyclicly hinged, the pieces in one of the 1-hexagons are tube-cyclicly hinged, and four of the pieces in the other 1-hexagon are flat-cyclicly hinged, as we see in Figure 18.11.

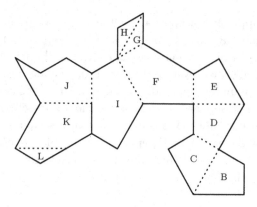

Figure 18.8.   Net for the $\sqrt{3}$-hexagon.

Even for triangles and hexagons, there is ample room to experiment with new combinations. I know of no previous dissection of a $\sqrt{3}$-triangle (three small triangles) and a 1-hexagon (six small triangles) to a 3-triangle (nine small triangles).   The 10-piece folding dissection in Figure 18.13 comes from converting the $\sqrt{3}$-triangle to a hexagonal net, converting the 1-hexagon to a net containing two hexagons, combining the two nets to give a tessellation element, and superposing it with the tessellation of 3-triangles. The $\sqrt{3}$-triangle has a cap-cyclic hinging. Indeed, it looks familiar, because we have seen the same cutting and hinging of the triangle in Figure 9.2.

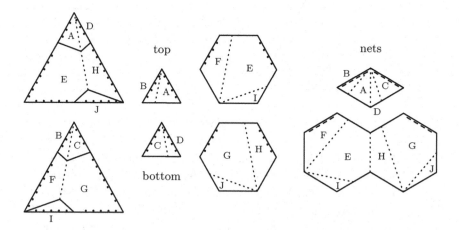

Figure 18.9.   Folding dissection of a 1-triangle and a 1-hexagon to a $\sqrt{7}$-triangle.

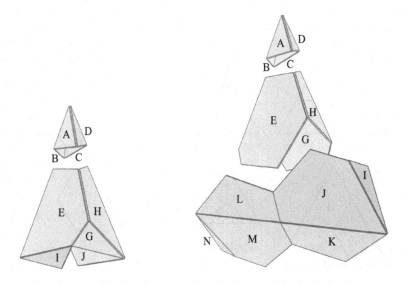

Figure 18.10. Forming √7-triangle.

Figure 18.11. Forming √13-triangle.

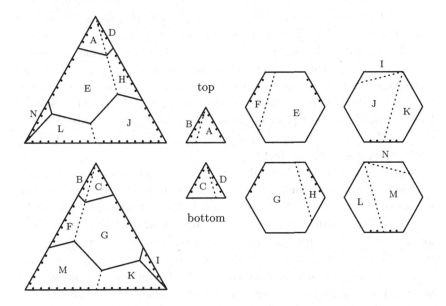

Figure 18.12. Folding a 1-triangle and two 1-hexagons to a √13-triangle. **(C)**

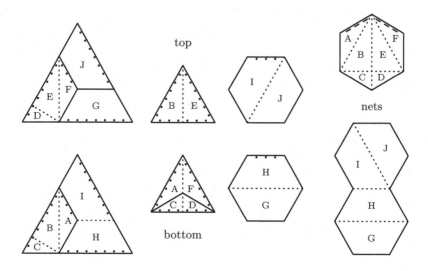

Figure 18.13. Folding a $\sqrt{3}$-triangle and a 1-hexagon to a 3-triangle.

Up to this point in the chapter, hexagrams have been sorely underrepresented figures. Clearly, this is an intolerable situation for which we will now seek redress. Alfred Varsady found a 5-piece unhingeable dissection of a 2-triangle and a 1-hexagram to a 4-triangle. The 11-piece folding dissection in Figure 18.14 uses the fact that the 1-hexagram and the 4-triangle are the same height. Much of the top level of the 4-triangle comes from the top level of the hexagram. Pieces D and G must fold down to the bottom level, necessitating that piece G drags piece F along for the ride. We fold the rest of the pieces to fill in as needed. Pieces A and I extend into both levels, as we see in Figure 18.15.

Another convenient relationship is of a 1-hexagon (six triangles) and a 1-hexagram (twelve triangles) to a $\sqrt{3}$-hexagon (eighteen triangles). There is a symmetric unhingeable dissection that uses just six pieces. Similarly, there is a simple symmetric folding dissection, but it has 20 pieces—quite a lot! I struggled to reduce the number of pieces, eventually getting it down to the 12 that we see in Figure 18.16. All symmetry now seems to be gone, and it is indeed difficult to verify that the folding actually works without making a model. Perspective views of the assemblages are in Figure 18.17.

A similar relationship is of a 2-hexagon (24 triangles) and a 1-hexagram (12 triangles) to a $\sqrt{3}$-hexagon (36 triangles). In contrast to the last

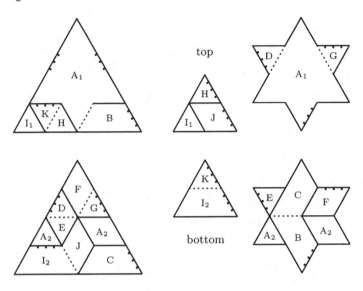

Figure 18.14. Folding a 2-triangle and a 1-hexagram to a 4-triangle.

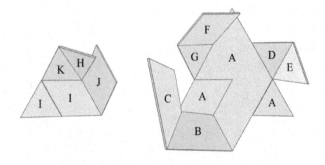

Figure 18.15. Assemblages for a 2-triangle and a 1-hexagram to a 4-triangle.

dissection, this one is inclusive of a beautiful symmetry. There are two swing-hingeable dissections that use just seven pieces. Both are lovely, with sixfold rotational symmetry. With some work, I have adapted some of the ideas from them to give the 23-piece folding dissection in Figure 18.18. In each of the smaller figures, I cut and unfold one of the two levels, using the unfolded pieces to fill in the points of the large hexagram. Compared with the previous dissection, there are a lot of pieces, and I tried hard to reduce the number. However, forming the points for the $\sqrt{3}$-hexagram

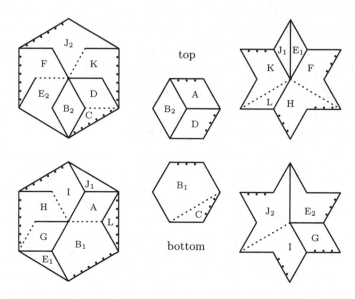

Figure 18.16. Folding a 1-hexagon and a 1-hexagram to a $\sqrt{3}$-hexagon.

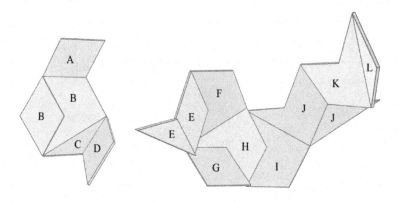

Figure 18.17. Assemblages for a 1-hexagon and a 1-hexagram to a $\sqrt{3}$-hexagon.

here seems more difficult than forming the sides of the $\sqrt{3}$-hexagon in the previous dissection. As consolation, we still have lovely symmetry, as we see in Figure 18.19.

Next are two dissections of questionable geometricity (geometric origin). I know of no previous dissection of a $\sqrt{19}$-triangle (19 triangles) and a 1-hexagon (six triangles) to a 5-triangle (25 triangles). The 11-piece folding

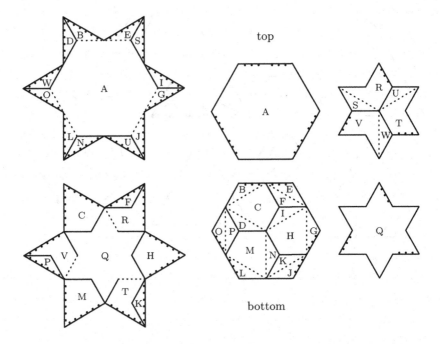

Figure 18.18. Folding a 2-hexagon and a 1-hexagram to a √3-hexagram.

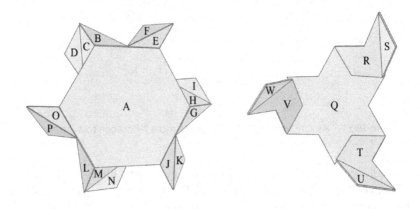

Figure 18.19. Assemblages for a 2-hexagon and a 1-hexagram to a √3-hexagram.

dissection in Figure 18.20 comes from converting the √19-triangle to a polygon bounded by line segments between points on a triangular lattice. I chose the polygon so that I can fold it to fill the right side of the 5-triangle.

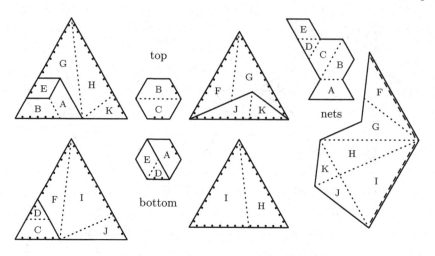

Figure 18.20. Folding a $\sqrt{19}$-triangle and a 1-hexagon to a 5-triangle.

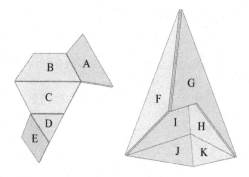

Figure 18.21. Assemblages for a $\sqrt{19}$-triangle and a 1-hexagon to a 5-triangle.

The challenge is then to cut and fold the hexagon to fill in the remaining space of the 5-triangle. As we can see in Figure 18.21, the pieces from the $\sqrt{19}$-triangle have a cap-cyclic and a flat-cyclic hinging.

Again, I know of no previous dissection of a $\sqrt{7}$-triangle (seven triangles) and a 1-hexagon (six triangles) to a $\sqrt{13}$-triangle (13 triangles). The 14-piece folding dissection in Figure 18.22 comes from converting the $\sqrt{7}$-triangle to a figure that fits against the hexagon and then superposing tessellations. The nets, the element from converting the $\sqrt{7}$-triangle, and the superposed tessellations are in Figure 18.23. The two little pieces, G

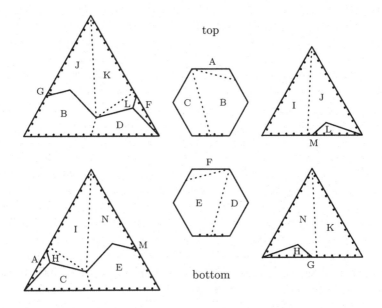

Figure 18.22. Folding a $\sqrt{7}$-triangle and a 1-hexagon to a $\sqrt{13}$-triangle.

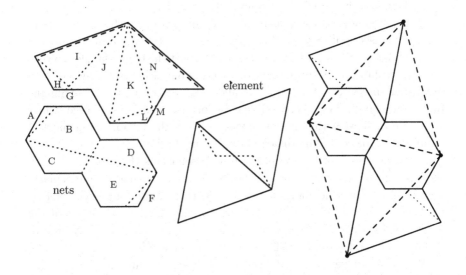

Figure 18.23. Nets, element, and superposition.

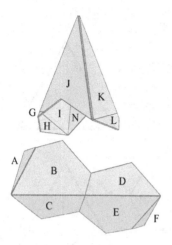

Figure 18.24. A first √13-triangle.

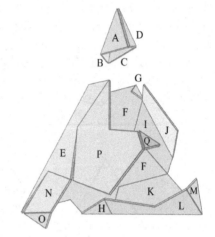

Figure 18.25. A second √13-triangle.

and M, are hinged to H and J and to L and N, respectively. Indeed, each participates in a different flat-cyclic hinging. We can see in Figure 18.24 that pieces I, J, K, and N are cap-cyclicly hinged.

What a wonderful assortment of cyclic hingings, which we also see in our next dissection! Since a hexagram contains twelve triangles and we can lay a large triangle whose area equals that of thirteen triangles so that its vertices fall on the gridpoints of a triangular grid, there should be a propitious dissection of a 1-triangle and a 1-hexagram to a √13-triangle. Indeed, we can find a 7-piece unhingeable one, as well as an 8-piece swing-hingeable dissection that is not hinge-snug. With this encouragement, I found the 17-piece folding dissection in Figure 18.26.

It too derives from a superposition of tessellations (Figure 18.27), with the one coming from the star and the small triangle being not particularly easy to find. Cut and fold out the star in the manner suggested in Figure 1.3. Then, fold out the small triangle, and place it to the left of the lower point of the star. The pieces in the 1-triangle are cap-cyclicly hinged, and pieces F, H, K, and L are flat-cyclicly hinged, as we see in Figure 18.25.

We have not yet seen a dissection involving all three of the triangle, the hexagon, and the hexagram. It is high time to celebrate our figures' differences. Counting small triangles, we verify that a 1-hexagram (twelve triangles) plus a 2-hexagon (24 triangles) gives a 6-triangle (36 triangles). It is not too hard to find the 14-piece rounded piano-hinged dissection in

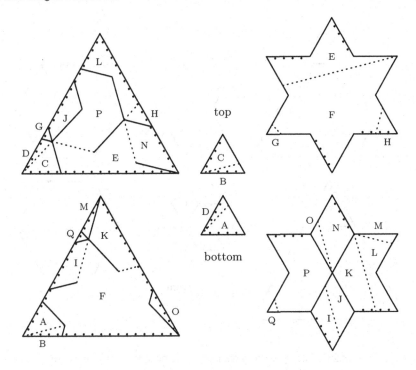

Figure 18.26. Folding a 1-triangle and a 1-hexagram to a √13-triangle.

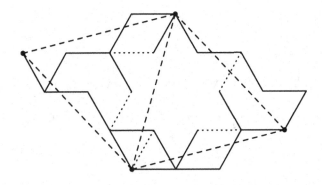

Figure 18.27. Tessellation for a 1-triangle and a 1-hexagram to a √13-triangle.

Figure 18.28. We simply fold the hexagram out into a one-level-thick 2-hexagon and then fold the top level of the given 2-hexagon to fill one level of the 6-triangle plus its corners on the other level, as we see in Figure 18.29.

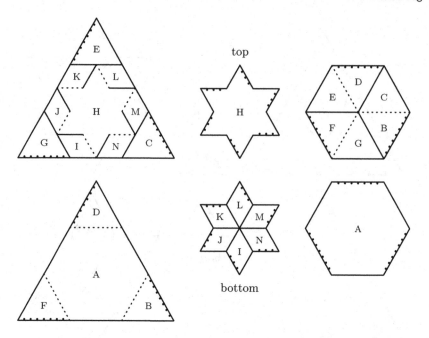

Figure 18.28. Folding a 1-hexagram and a 2-hexagon to a 6-triangle. (C)

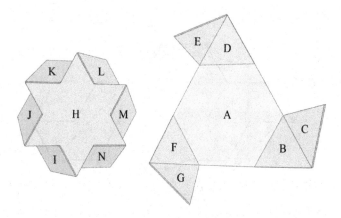

Figure 18.29. Assemblages for a 1-hexagram and a 2-hexagon to a 6-triangle.

There is an even easier folding dissection involving a triangle, a hexagon, and a hexagram, which I leave as a puzzle.

Puzzle 18.2. *Find an 11-piece folding dissection of a 1-hexagram and a $2\sqrt{3}$-triangle to a 2-hexagon.*

By now we have mixed things up about as well as we can for triangles, hexagons, and hexagrams. In the next chapter, we will broaden our horizons still further and achieve an even greater diversity by including polygons and stars with five or ten vertices, as well as other polygons and stars.

Chapter 19

Pot Luck

I sat at the dining-room table that cold, dark, February night, constructing a flimsy paper model on which to test my brainstorm from earlier in the evening. It had seemed so nifty, yet so improbable. Every hinge in the 10-pointed star was to participate in one of five saddle-cyclic hingings, turning the figure completely inside-out. Surely the pieces would obstruct each other, when I attempted to transform the star to a "holey" decagon. It would be too pretty otherwise.

As I snipped with my pair of scissors, my daughter finally located me to say good night. I was in an unaccustomed corner of the house, doing the unexpected. Later, as I was taping the tiny pieces together, my wife looked in and observed that my model was too small. I countered that if it didn't work, I wouldn't want a large model to remind me of my disappointment. When I finished the assembly, I put it to the test. To my amazement, it seemed to work! Over and under, around and back, I checked the delicate mechanical structure several times, just to make sure. After being subjected to a demonstration, my wife even registered that rare (for her) show of astonishment.

As an inventor, one must consider all manner of inspired ideas and keep one's mind open to crazy combinations of things. Since it's "pot luck," things don't always work. But when they do, it's terrific! And when they elegantly capture the essence of some nifty idea, it's heaven! We'll explore unusual unions of polygons in this chapter, working our way through a gallery of wild wonders, heading towards that 10-pointed one and some other surprises, too.

Let's start with a simple beauty, a 6-piece unhingeable dissection of two pentagrams and two pentagons to a decagon, which Ernest Freese (1957b) found. He cut each star into two pieces and rearranged them to fit around

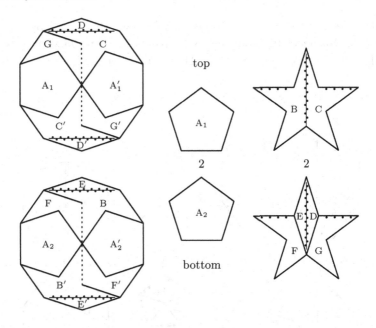

Figure 19.1. Two pentagrams and two pentagons to a decagon.

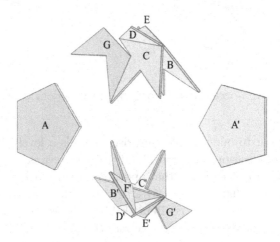

Figure 19.2. View of two pentagrams and two pentagons.

the uncut pentagons. To find a folding dissection, I follow Freese's lead of not cutting the pentagons. Of course, cutting the pentagrams is a bit trickier. To get the 14-piece dissection in Figure 19.1, start by flipping piece F next to piece B, and similarly for pieces G and C. The key is to use a leaf-cyclic hinging on pieces B, C, D, and E, as we see in Figure 19.2.

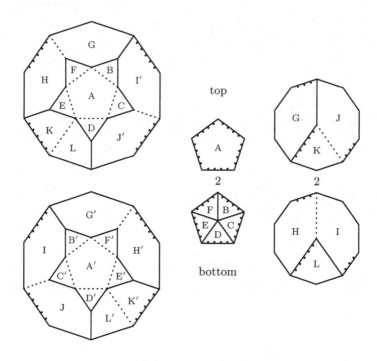

Figure 19.3. Two 1-decagons and two $(2\sin(\pi/5))$-pentagons to a ϕ-decagon.

Moving closer to our ultimate surprise, there is a lovely 12-piece dissection of two 1-decagons and two $(2\sin(\pi/5))$-pentagons to a ϕ-decagon, which is related to a dissection that appeared in the circa 1300 *Interlocks* manuscript. Recall that $\phi = (1+\sqrt{5})/2 \approx 1.618$ is the golden ratio. We can show that $2\sin(\pi/5) = \sqrt{4-\phi^2}$. Remarkably, there is a direct adaptation of this dissection of seven centuries ago into the 24-piece folding dissection in Figure 19.3. You can see the pieces as they fold in Figure 19.4. Can you identify the 12-piece dissection that is inspired from so long ago? It is swing-hingeable.

M. N. Deshpande (2002) identified essentially the same relationship behind the dissection in the *Interlocks* manuscript, namely that we can dissect a decagon into twenty isosceles triangles and the pseudostar $\{5/2.5\}$. He

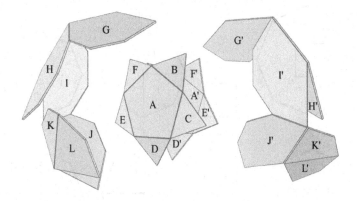

Figure 19.4. View of two pentagons and two decagons.

also described how to dissect (analogously) an octagon into twelve congruent isosceles triangles and a small square. Each isosceles triangle has an apex of 45°, and there are eight such triangles in a smaller octagon.

I have found two different dissections based on this octagon relationship, the first of which is of two 1-octagons to three $(2\sin(\pi/8))$-octagons and two $(2\sin(\pi/8))$-squares. Note that

$$2\sin(\pi/8) = \sqrt{2 - \sqrt{2}}.$$

Discovering a folding dissection for that one proved a challenge, but I was able to find the 24-piece dissection in Figure 19.5. I cut each of the large octagons identically, and I also cut two of the three small octagons identically. The cuts and hinges in the third small octagon are rotationally symmetric, rotating from top to bottom. How surprising that pieces F, G, H, and I are hinged cap-cyclicly, as we see in Figure 19.6.

My second dissection for the octagon relationship is of three 1-{8/3}s and four $(2\sin(\pi/8))$-squares to four 1-octagons. It is possible to find a 22-piece unhinged dissection and a 28-piece swing-hinged dissection. When I rolled up my sleeves, I found the 37-piece rounded piano-hinged dissection in Figure 19.7. I cut three of the octagons identically, with the cuts being almost symmetric with respect to top and bottom, except that piece C corresponds to pieces L and M. The third of the {8/3}s is actually different from the one that I show in the figure. The portion filled by pieces N, O, and P should be filled by pieces Q, R, S, T, U, and V. Note that I have not shown the four squares separately in Figure 19.7.

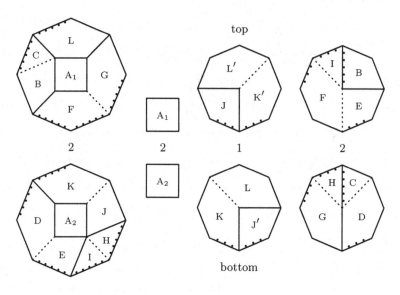

Figure 19.5.   Two octagons to three octagons and two squares.   (C)

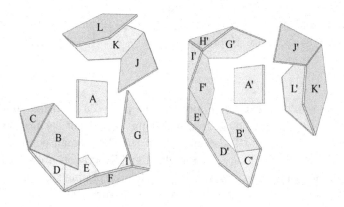

Figure 19.6.   View of three octagons and two squares.

At first, I also used a hinge between pieces H and I, so that the dissection had a saddle-cyclic hinging of pieces F, G, H, and I. However, after examining the next dissection carefully, I realized that this additional hinge would not work. Yet, there is a cap-cyclic hinging among pieces R, S, T, and U. This appears in Figure 19.8, where we see a view of the assemblages for two of the three {8/3}s. Please do not try to merge piece M

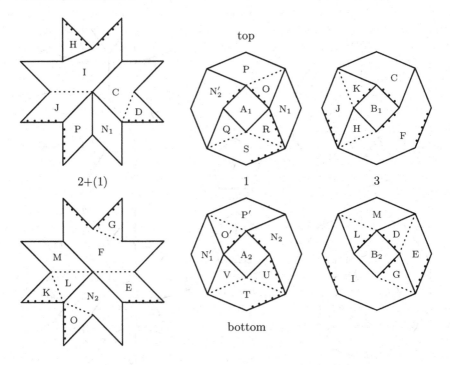

2+(1)   1   3

bottom

Figure 19.7. Rounded three {8/3}s and four squares to four octagons.

Figure 19.8. View of two of the three 1-{8/3}s.

with piece L. I made this split so that we could fold from the star to the octagon without bending any pieces.

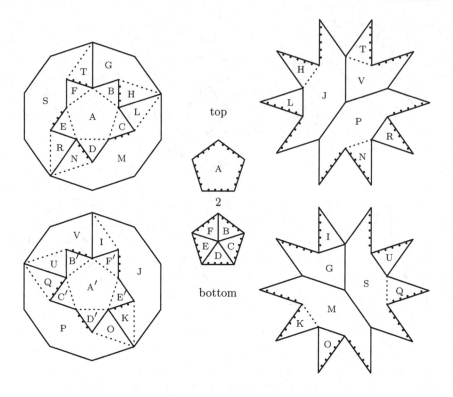

Figure 19.9.   Rounded {10/4} and two pentagons to a decagon. **(W)**

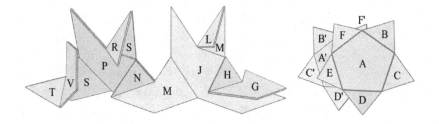

Figure 19.10.   View of assemblages for a φ-decagon.

Having discovered the dissections in Figures 19.5 and 19.7 and then recognized the connection with Figure 19.3, it finally hit me that I should try to aim for a {10/4} rather than the two decagons in Figure 19.3. I would not have had much hope, except that the saddle-cyclic hinging that

I thought that I could use in Figure 19.7 seemed to give the needed leverage. The "bridge piece" between two points of the 10-pointed star would have two "nubs," each of which would participate in a saddle-cyclic hinging with another bridge piece.

Although the saddle-cyclic hingings seemed to work fine with thin flexible materials such as paper and cardboard, they did not work properly with a truly rigid material. The problem is in having angles greater than 180° involved in the saddle cycle. After experimentation, I found that removing one piano hinge from each cycle made a rounded piano-hinged dissection possible. The resulting 28-piece dissection is in Figure 19.9. You can get some appreciation of the complexity of this dissection in the perspective view of Figure 19.10. Even without the desired five saddle-cyclic hingings—(G, H, I, J), (J, K, L, M), (M, N, O, P), (P, Q, R, S), and

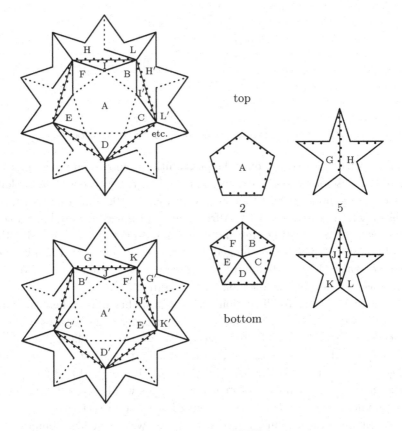

Figure 19.11. Five 1-pentagrams and two $\sqrt{4-\phi^2}$-pentagons to a 1-{10/3}.

(S, T, U, V)—it is a real beauty! And if you make the pieces really thin, then you can probably fool yourself, like I did, with the saddle-cyclic hingings.

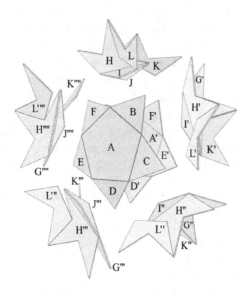

Figure 19.12.   View of five pentagrams and two pentagons.

Have we exhausted not only the possibilities from the *Interlocks* manuscript, but also ourselves? Certainly not the latter, because we can exploit the leaf-cyclic hinging trick that we saw in Figures 19.1 and 19.2 to attack a dissection of Alfred Varsady. Alfred found a 16-piece (swing-hingeable) dissection of five 1-pentagrams and two $\sqrt{4-\phi^2}$-pentagons to a 1-{10/3}. To obtain my 42-piece piano-hinged dissection (Figure 19.11), I handle the pentagons as in Figure 19.3 and the pentagrams as in Figure 19.1. We see a perspective view of the assemblages in Figure 19.12.

The original Varsady dissection with which we began is rather straightforward, and this should encourage us to consider other relatively easy dissections. Jean Bauer (1999) noticed that we can cut the points off a $\{p/q\}$ star and fit them into the cavities of a $\{p/(q+1)\}$ star of the same sidelength, forming two congruent polygons $\{p\}$. If the $\{p/q\}$ and $\{p/(q+1)\}$ have sidelength 1, then the sidelength of the $\{p\}$s will be $2\cos(2\pi/p)$. This leads immediately to a $(p+2)$-piece swing-hingeable dissection.

When we turn to piano-hinged dissections, we are at first tempted to cut each star to separate its levels and then perform the above operation,

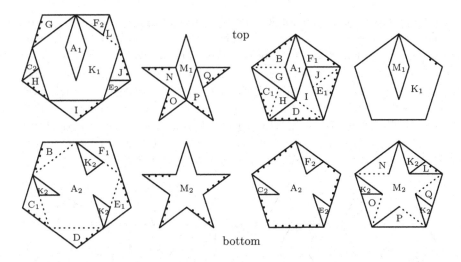

top

bottom

Figure 19.13.   A 2-pentagon and a 1-pentagram to two $(2\cos(2\pi/5))$-pentagons.

yielding a $(2p+4)$-piece dissection that appears to be piano-hinged. However, this approach hinges the pieces into four assemblages, which is one more than the minimum of three assemblages. We simply must do better!

So, these dissection puzzles become more interesting than they first appeared. What clever contrivances do we need in order to create just three assemblages? We can use two tricks that we saw earlier, in Figure 8.9. In that dissection, there was a piece of double thickness near the center of one

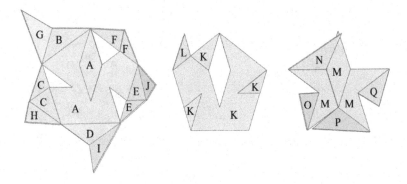

Figure 19.14.   View of pentagon and pentagram to two pentagons.

of the squares. To accommodate that piece when forming the other square, there was a piece with a hole in it. This all sounds reasonable but does not prepare us for the wacky weirdness of the shapes to which we will resort.

Our first surprise is that even though there is only one type of 5-pointed star, we can start with $p = 5$. We take $q = 1$ and note that a "star" with each vertex connected to the next one is simply a regular polygon. The sidelength of that regular $\{p/1\}$ polygon will be 2 when the sidelength of the $\{p/2\}$ star is 1. Yet, even converting a regular pentagon and a pentagram to a pair of pentagons, using only three piano-hinged assemblages, is not

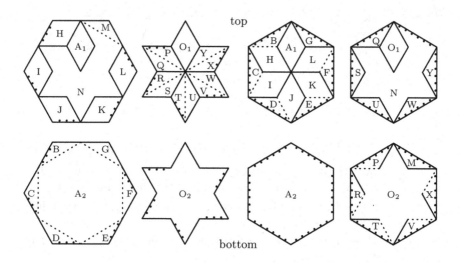

Figure 19.15. A 2-hexagon and a 1-hexagram to two $\sqrt{3}$-hexagons.

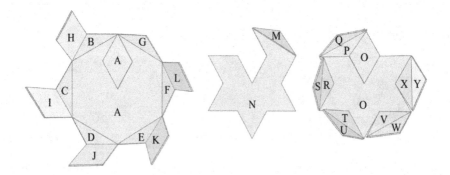

Figure 19.16. View of a 2-hexagon and a 1-hexagram to two $\sqrt{3}$-hexagons.

so easy. Our second surprise is that we can find a 17-piece piano-hinged dissection that fits our requirements.

We cut just five pieces out of one of the $(2\cos(2\pi/5))$-pentagons to form the pentagram. What remains, together with the other $(2\cos(2\pi/5))$-pentagon, must form the 2-pentagon. We then juggle various pieces around so that piece K has three projections into its other level and four pieces (A, C, E, and F) from the second $(2\cos(2\pi/5))$-pentagon are also on both

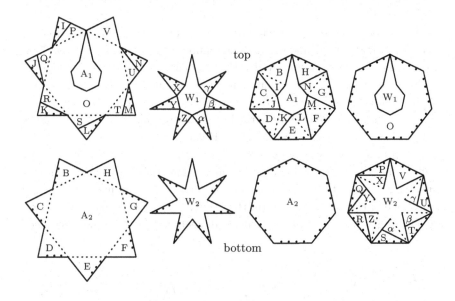

Figure 19.17. A 1-{7/2} and a 1-{7/3} to two $(2\cos(2\pi/7))$-heptagons.

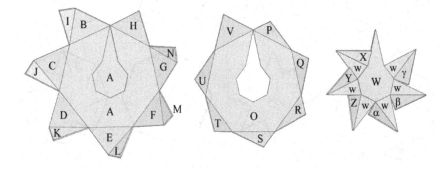

Figure 19.18. View of a 1-{7/2} and a 1-{7/3} to two $(2\cos(2\pi/7))$-heptagons.

levels. I discovered this interlocking structure in Figure 19.13 by trial and error. As we can see in Figure 19.14, this dissection is quite decidedly rounded.

For 6-pointed figures, we can dissect a 2-hexagon plus a 1-hexagram to two $\sqrt{3}$-hexagons. For a piano-hinged dissection, I haven't done nearly as well as far as minimizing the number of pieces, finding a 25-piece dissection (Figure 19.15). The problem is that I have not found an economical way to fill in the bottom level of the hexagram. We see a perspective view in Figure 19.16.

When we come to 7-pointed figures, we can involve all of the 7-pointed figures if we take $p = 7$ and $q = 2$. The number of pieces in Figure 19.17, at 29, is somewhat large. I have folded the top of six of the points of the $\{7/3\}$ down to the bottom level. What remains to be filled in when forming one of the heptagons seems to dictate how to cut all of the remaining pieces. This is perhaps the most symmetrical dissection in this series of four piano-hinged dissections, as we see in Figure 19.18.

With 8-pointed objects ($p = 8$ and $q = 2$), we can once again do very well. I have just 26 pieces, only six more than the straightforward not-fully-hinged dissection that has four assemblages. I cut just eight pieces out of one of the octagons to form the $\{8/3\}$. I base the 26-piece dissection in Figure 19.19 on a simpler 27-piece dissection. In that one, the piece

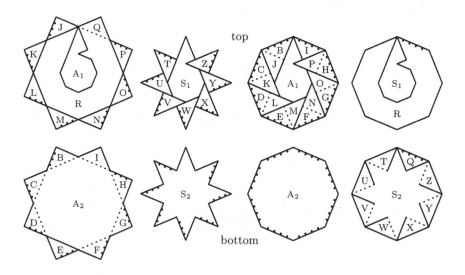

Figure 19.19. A 1-$\{8/2\}$ and a 1-$\{8/3\}$ to two $\sqrt{2}$-octagons. (C)

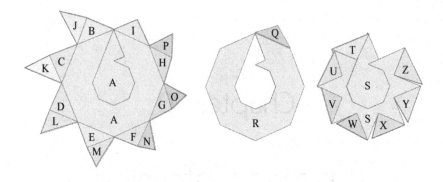

Figure 19.20.   Perspective view of a 1-{8/2} and a 1-{8/3} to two $\sqrt{2}$-octagons.

corresponding to piece S has an upper portion in the form of a small octagon. We can save a piece by gluing an isosceles right triangle onto that small octagon and swapping right triangles around as necessary. The three assemblages flap to life in Figure 19.20.

Well, now you have just about reached the end of this book. Will your family soon find you sitting at your own dining-room table on a cold, dark, winter evening? Will you be measuring out the pieces on fine cardboard for one of these diabolical dissections and then swinging your scissors into action? Or maybe you will have been beset by your own brainstorm, boggled and bothered into producing your own prototype. Will you find yourself feverishly folding out your design, caught up in the suspense, wondering whether it will unfold the way you would like? Let's hope so. Have a wild one!

Chapter 20

Puzzles Unfolded

This chapter supplies solutions to the numbered puzzles throughout the book. The number of the solution coincides with the number of the puzzle.

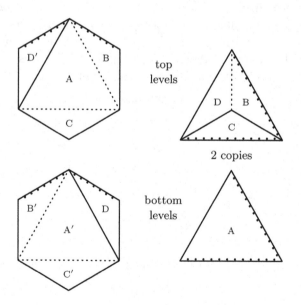

Solution 1.1. A different piano-hinging for a hexagon to two triangles.

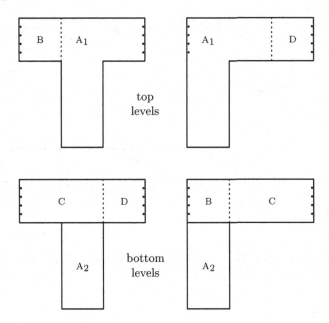

Solution 2.1.   A 4-piece tube-cyclicly hinged T-pentomino to a V-pentomino.

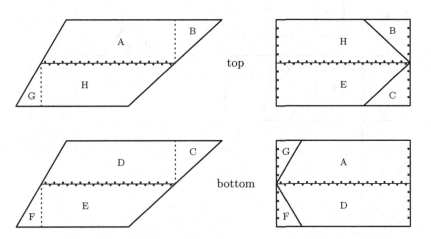

Solution 5.1.   Leaf-cyclic hinging of a trapezoid to a rectangle.

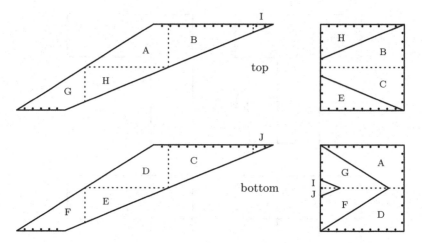

Solution 5.2. Folding trapezoid with an uncooperative side to a rectangle.

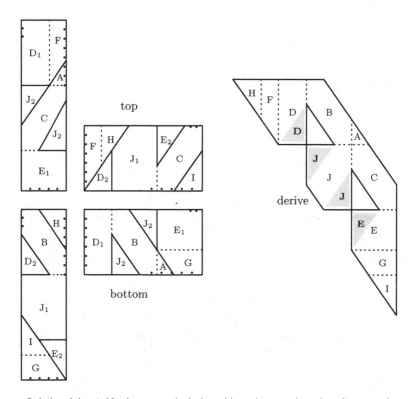

Solution 6.1. A 10-piece rounded piano-hinged rectangles when $2 < \alpha < 3$.

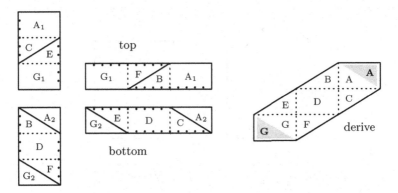

Solution 6.2. 7-Piece piano-hinged (l, w)-rectangle to a ($3l$, $w/3$)-rectangle.

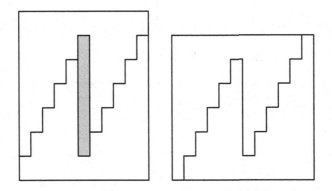

Solution 7.1. Unhinged (11 × 14)-rectangle with 1 × 10 hole to a square.

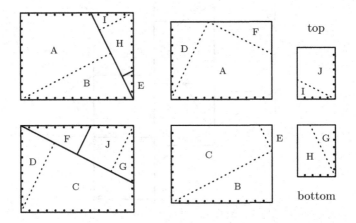

Solution 8.1. Folding dissection of two similar rectangles to one.

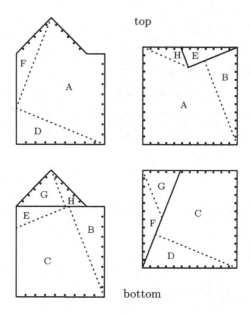

top

bottom

Solution 8.2.   Folding dissection of small-roof house to a square.

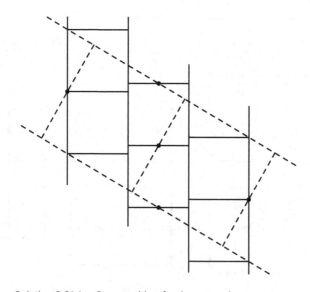

Solution 8.3(a).   Crossposition for three equal squares to one.

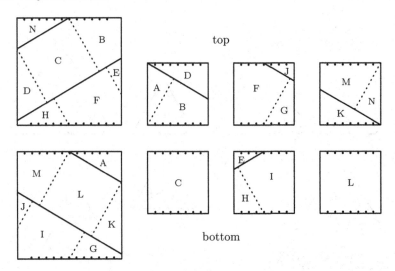

Solution 8.3(b). Folding dissection of three equal squares to one.

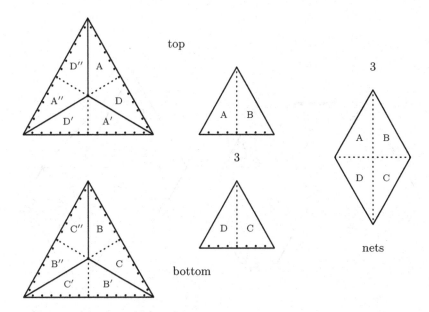

Solution 9.1. Flat-cyclic-hinged dissection of three equal triangles to one.

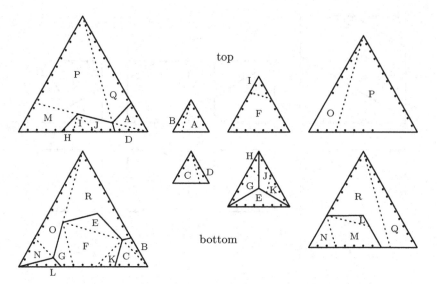

Solution 9.2(a). Folding dissection of triangles for $1^2 + (\sqrt{3})^2 + 3^2 = (\sqrt{13})^2$.

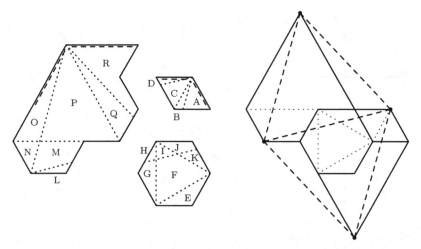

Solution 9.2(b). Superposition and nets: triangles for $1^2 + (\sqrt{3})^2 + 3^2 = (\sqrt{13})^2$.

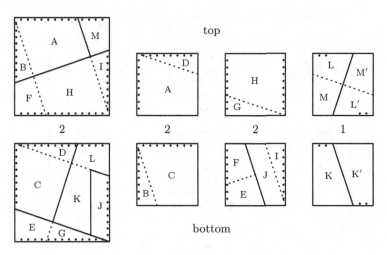

Solution 10.1.   Tessellation-based five squares to two.

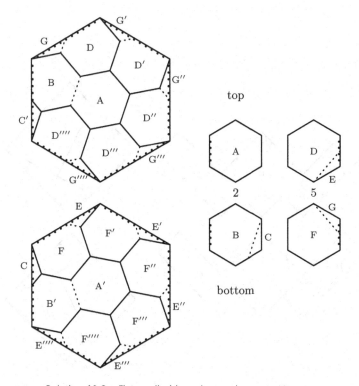

Solution 10.2.   Flat-cyclic-hinged seven hexagons to one.

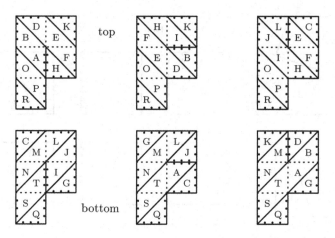

Solution 11.1.   Three of four distinct ways to form the P-pentomino.

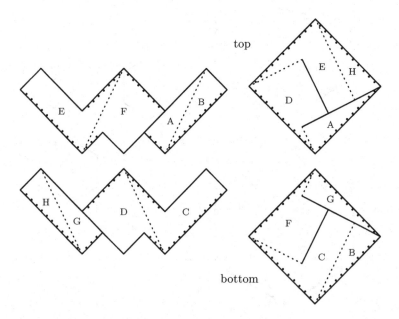

Solution 12.1.   Decomino dissection analogous to W-pentomino dissection.

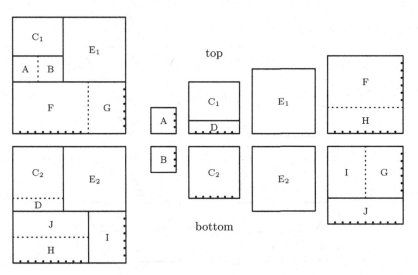

Solution 13.1.   Piano-hinged dissection of squares for $2^2 + 4^2 + 5^2 + 6^2 = 9^2$.

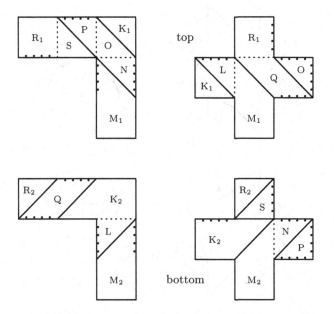

Solution 15.1.   Piano-hinged L-pentomino to a Greek Cross.

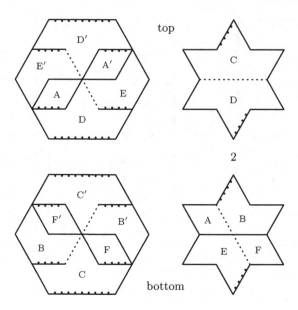

top

2

bottom

Solution 16.1.   Folding dissection of two hexagrams to a hexagon.

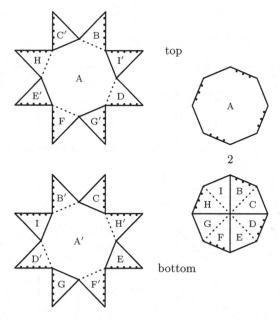

top

2

bottom

Solution 16.2.   Different fold-hinging of an {8/3} to two octagons.

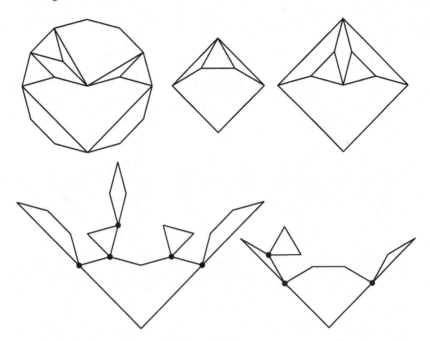

Solution M5.1. Hinge-snug swing-hinged {12} to 1- and $\sqrt{2}$-squares.

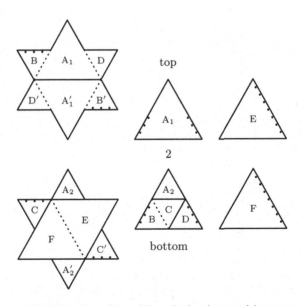

Solution 18.1. Piano-hinged three 2-triangles to a 1-hexagram.

top

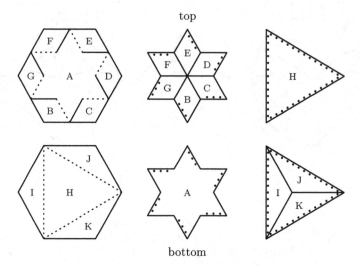

bottom

Solution 18.2. Folding 1-hexagram and $2\sqrt{3}$-triangle to a 2-hexagon.

Afterword

Taking a broad view, folding almost certainly predates recorded history. Would early humans have folded their animal skins prior to a migration? Might they have wrapped a valuable gift in a large leaf to surprise the recipient? Once people had learned to weave cloth, they no doubt stored it in folded form. Over time, such a fundamental activity would necessarily find its way into artistic expression and recreation. Folding has entered into our culture in manifestations as different as origami (see, for example, Engel (1989)), napkin folding (Harsdörffer (1657)), magic (Houdini (1922)), folding puzzles (Slocum and Botermans (1986)), flexagons (Pook (2003)), and even pop-up books (Diaz and Carter (1999)). Origami, probably the most popular of these activities, has benefited in recent years from an influx of mathematical methods (Lang (2003)).

A deputy tax collector in Madras, India, Tandalam Sundara Row (1893) reproduced Hanumanta Rau's description of how to fold a triangular piece of paper to illustrate the property that the three angles of a triangle sum to 180°. A Professor Lundgren (see Hoffmann (1900)) also described the same construction. Henry Martyn Cundy and Arthur Percy Rollett (1952) placed the same model in the context of other hinged dissections and illustrated the folding quite clearly. Alton Olson (1975) reviewed examples of paper folding that illustrate mathematics. Henry Petroski (2005) waxed nostalgic as he surveyed various techniques that paperboys used for folding the newspapers that they tossed onto subscribers' stoops and porches.

Howard Eves (1969) recounted the story of how the seventeenth-century French mathematician Blaise Pascal discovered as a boy the fact that the sum of the angles of a triangle is equal to a straight angle. Eves claimed that Pascal's sister mentioned that the fact was discovered by some process of folding, with Eves suggesting Sundaram Row's demonstration as one possibility for this folding. Unfortunately, Eves's version of the story appears to be at variance with every other recounting. Nowhere else is there the suggestion that Pascal considered folding.

Pop-up artists have considered constructions that are very similar to vertex-cyclic hingings. In particular, David Carter and James Diaz (1999) describe a "180° angle fold pyramid" and a "180° angle fold platform" that correspond closely to a cap-cyclicly hinged set of pieces and a saddle-cyclicly hinged set of pieces, respectively. Of course, since they discuss pop-up books, their constructions are flat in only one of two configurations.

Jin Akiyama and Gisaku Nakamura (1999) also exploited a piano-hinged paradigm, from the point of view of turning inwards every surface of a dissected and piano-hinged prism. Their constructions include two right triangles to an irregular pentagon, a convex quadrilateral to two parallelograms, and two right triangles to an isosceles triangle. They also identified a property similar to the inside-out property with respect to (two-dimensional) swing-hinged dissections (2000a). They required their dissections, which they called Dudeney dissections, to have every edge be in the boundary of one figure or the boundary of the other figure, but not both. For three-dimensional figures, they required the boundaries, which are surfaces for three-dimensional figures, to have the same sort of property (2000b).

David Huffman (1976) identified a model and analyzed topological properties in a fashion consistent with, but different from, what I give in Chapter 2. Vertex-cyclic hingings are reminiscent of cones with varying cone angles, which were discussed by David Henderson (1996). Tube-cyclic hingings share similarities with several classes of folding mechanisms explored by the Cambridge mathematician Geoffrey Bennett (1905).

When I show a perspective view of a piano-hinged assemblage that has a cap-cycle, I will almost always show it in a configuration in which the dihedral angles between consecutive pieces around the cap-cycle are equal. (The *dihedral angle* between two pieces that share an edge is the angle between them in the plane that cuts both pieces perpendicular to the edge.) Suppose that $\alpha \leq \sigma - \alpha$. Then, the pieces with angle α at the vertex will be separated from each other by an angle of $2\arcsin(\sin((\sigma - \alpha)/2)/\cos(\alpha/2))$, and the other two pieces will be separated by an angle of $2\arcsin(\sin(\alpha/2)/\cos((\sigma - \alpha)/2))$. (Those who have studied trigonometry may recall that the *arcsine*, or *inverse sine*, of a value r is the angle α such that $\sin(\alpha) = r$.) In Figure 2.7, for example, opposite pieces (such as A and C) in the cap-cycles will be at an angle of $2\arcsin(\sin 22.5°/\cos 22.5°) \approx 48.94°$ with each other.

After writing Folderol 1, I submitted and published it as a journal article (2003a). Subsequently, the article was awarded the Mathematical Association of America's George Pólya Award, given annually for an article of "expository excellence published in *The College Mathematics Journal.*" I have also excerpted portions of the book in (2003b), (2005a), and (2005b).

School Superintendent W. W. Ross (1891) described a way to fold paper so as to demonstrate that a trapezoid has the same area as a rectangle of the same height and length equal to half the sum of the trapezoid's two bases. Ross's demonstration was not a piano-hinged dissection as defined here, and yet there are vague similarities between the figure on his page 10 and my Figure 5.1.

Aydin Sayili (1958) gave the Arabic text, Turkish translation, explanation, and analysis of Thābit's dissection. He also gave a discussion in English (1960). Lindgren's 13-piece dissection of seven triangles to one was anticipated by Jan G.-Mikusiński (1946) and Hugo Steinhaus (1950).

My apologies to whichever screenwriter of the 1939 film version of *The Wizard of Oz* penned the phrase "lions, and tigers, and bears—oh my!"

In Manuscript 4, I used the names of the polygons as determined by John Conway and Antreas Hatzipolakis in consultation with several other scholars. (See "Naming Polygons and Polyhedra" in the "Ask Dr. Math" column on the Math Forum at Drexel University.[‡]) Gavin Theobald's terminology varies somewhat: undecagon, tridecagon, tetradecagon, pentadecagon, hexadecagon, heptadecagon, octadecagon, enneadecagon, and tetra-icosagon. Conway and Hatzipolakis gave tetradecagon, pentadecagon, and hexadecagon as alternative forms.

Freese (1930b) gave data on regular polygons up to 24 sides, including the area, the radius of the circumscribing circle, and the radius of the inscribing circle. He wrote:

> Incidentally, you'll probably wonder how I procured the *names* of the last twelve polygons listed in Table 4. Well, I *guessed* at the final *four*—but the others are "authentic." I'll admit, however, that it's much more comprehensive to say "*24 sides*" than it is to warble "tetrabisdecagon." Maybe I'm wrong!

The quote from Chief Inspector Dreyfus (portrayed by Herbert Lom) comes in the fifth of the pink panther movies, *The Pink Panther Strikes Again*, directed by Blake Edwards, 1976. The slogan originated with the French pharmacist Emile Coué (1857–1926) who instructed his patients to repeat the sentence. This autosuggestion was very popular in the 1920s.

Greg Blonder (2003) identified (independently) what is essentially a dissection of a one-level-thick $\{10/4\}$ to a two-level-thick decagon.

Jean Bauer, Jean-Phillipe Lebet, and Stéphane Pécant (2001) have produced a lovely CD-book that expands greatly on the earlier work of Jean Bauer (1999).

[‡] http://mathforum.org/dr.math/faq/faq.polygon.names.html

Bibliography

Abu'l-Wafā' al-Būzjānī. *Kitāb fīmā yahtāju al-sāni' min a' māl al-handasa* (*On the Geometric Constructions Necessary for the Artisan*). Mashdad: Imam Riza 37, copied in the late 10th or the early 11th century. Persian manuscript.

Akiyama, Jin and Gisaku Nakamura (1999). Dudeney dissections of polyhedrons II, –layered type. In J. Akiyama, M. Kano, and M. Urabe (Eds.), *Discrete and Computational Geometry, Japanese Conference, JCDCG'99*. In Japanese.

Akiyama, Jin and Gisaku Nakamura (2000a). Dudeney dissection of polygons. In J. Akiyama, M. Kano, and M. Urabe (Eds.), *Discrete and Computational Geometry, Japanese Conference, JCDCG'98, Lecture Notes in Computer Science*, Volume 1763, pp. 14–29. Springer Verlag.

Akiyama, Jin and Gisaku Nakamura (2000b). Transformable solids exhibition. 32-page color catalogue.

Anonymous. *Fī tadākhul al-ashkāl al-mutashābiha aw al-mutawāfiqa* (*Interlocks of Similar or Complementary Figures*). Paris: Bibliothèque Nationale, ancien fonds. Persan 169, ff. 180r–199v.

Bauer, Jean (1999). Polygons' combinations. PowerPoint slide presentation.

Bauer, Jean, Jean-Phillipe Lebet, and Stéphane Pécant (2001). *Au delà du nombre d'or (Beyond the Golden Number)*. Neuchatel, Switzerland: Trigam. Electronic book on a CD.

Bennett, G. T. (1905). The parallel motion of Sarrut and some allied mechanisms. *London, Edinburgh, and Dublin Philosophical Magazine and Journal of Science* 9(54), 803–810.

Berloquin, Pierre (*Le Monde*). En toute logique. *Le Monde*. Column in the fortnightly section: des Sciences et des Techniques. (1974a): May 8, p. 22; (1974b): May 22, p. 20.

Blonder, Greg E. (2003). Interconvertible soft articles. U.S. Patent Application 20030216103. Filed 2002. See Figures 5A, 5B, 5C.

Bradley, H. C. (1930). Problem 3048. *American Mathematical Monthly* 37, 158–159.

Brodie, B. (1884). Superposition. *Knowledge* 5(135), 399.

Bruyr, Donald L. (1963). *Geometrical Models and Demonstrations*. Portland, Maine: J. Weston Walch.

Cardano, Girolamo (1663). *Hieronymi Cardani Mediolanensis, philosophi ac medici celeberrimi*, Volume III: De Rerum Varietate. Lugdani, Sumptibus Ioannis Antonii Hugvetan & Marci Antonii Ravaud. Originally published in 1557. Other titles: Works. Opera omnia. See p. 248 (mislabeled p. 348).

Catalan, Eugène (1873). *Géométrie Élémentaire* (fifth ed.). Paris: Dunod. See p. 194.

Cundy, H. M. and C. D. Langford (1960). On the dissection of a regular polygon into n equal and similar parts. *Mathematical Gazette* 44, 46.

Cundy, H. Martyn and A. P. Rollett (1952). *Mathematical Models*. Oxford.

Cuoco, Al (2000). Meta-problems in mathematics. *College Mathematics Journal* 31(5), 373–378.

de Fermat, Pierre (1891). *Œuvres de Fermat*. Paris: Gauthier-Villars et fils. See pages 140 and 149 in volume 1.

Demaine, Erik D., Martin L. Demaine, David Eppstein, Greg N. Frederickson, and Erich Friedman (2005). Hinged dissection of polyominoes and polyforms. *Computational Geometry: Theory and Applications* 31(3), 237–262.

Deshpande, M. N. (2002, July). A geometric dissection problem. *Resonance: Journal of Science Education* 7(7), 91.

Diaz, James and David Carter (1999). *Elements of Pop Up: A Pop Up Book for Aspiring Paper Engineers*. New York: Little Simon.

Dodge, Wally and Steve Viktora (2002). Thinking out of the box: A problem. Presented at the National Council of Teachers of Mathematics 80th Annual Meeting, Las Vegas, NV, April.

Dudeney, Henry Ernest (1907). *The Canterbury Puzzles and Other Curious Problems*. London: W. Heinemann. Revised edition printed by Dover Publications in 1958.

Dudeney, Henry Ernest (1917). *Amusements in Mathematics*. London: Thomas Nelson and Sons. Revised edition printed by Dover Publications, 1958.

Dudeney, Henry Ernest (1926). *Modern Puzzles and How to Solve Them*. London: C. Arthur Pearson.

Dudeney, Henry Ernest (1931). *Puzzles and Curious Problems*. London: Thomas Nelson and Sons. See problem 215.

Dudeney, Henry E. (*Cassell*). The puzzle realm. Column in *Cassell's Magazine*, 1908–1909. (1908a): September, p. 430.

Dudeney, Henry E. (*Dispatch*). Puzzles and prizes. Column in the *Weekly Dispatch*, April 19, 1896–Dec. 26, 1903. (1900a): August 26; (1902a): April 20; (1902b): May 4; (1903a): Sept. 13; (1903b): Nov. 29; (1903c): Dec. 13.

Dudeney, Henry E. (*Strand*). Perplexities. Monthly puzzle column in *The Strand Magazine* from 1910 to 1930. vols. 39–79. (1920a): vol. 59, p. 204; (1920b): vol. 59, p. 304; (1926a): vol. 72, p. 103.

Duemmel, James (1989). From calculus to number theory. *American Mathematical Monthly* 96(2), 140–3.

Dundas, Kay (1984). To build a better box. *College Mathematics Journal* 15(1), 30–6.

Elliott, C. S. (1982–1983). Some new geometric dissections. *Journal of Recreational Mathematics* 15(1), 19–27.

Elliott, C. S. (1985–1986). Some more geometric dissections. *Journal of Recreational Mathematics* 18(1), 9–16.

Engel, Peter (1989). *Folding the Universe: Origami from Angelfish to Zen*. New York: Vintage.

Essebaggers, Jan and Ivan Moscovich (1994). Triangle hinged puzzle. European Patent EP0584883. Filed 1993.

Esser, III, William L. (1985). Jewelry and the like adapted to define a plurality of objects or shapes. U.S. Patent 4,542,631. Filed 1983.

Eves, Howard W. (1969). *In Mathematical Circles*. Boston: Prindle, Weber & Schmidt.

Frederickson, Greg N. (1972a). Appendix H: Eight years after. In *Recreational Problems in Geometric Dissections and How to Solve Them*, by Harry Lindgren. New York: Dover Publications.

Frederickson, Greg N. (1972b). Several star dissections. *Journal of Recreational Mathematics* 5(1), 22–26.

Frederickson, Greg N. (1974). More geometric dissections. *Journal of Recreational Mathematics* 7(3), 206–212.

Frederickson, Greg N. (1997). *Dissections: Plane & Fancy*. New York: Cambridge University Press.

Frederickson, Greg N. (2002). *Hinged Dissections: Swinging & Twisting*. New York: Cambridge University Press.

Frederickson, Greg N. (2003a). A new wrinkle on an old folding problem. *College Mathematics Journal* 34(4), 258–263.

Frederickson, Greg N. (2003b). Piano-hinged dissections: now let's fold. In J. Akiyama and M. Kano (Eds.), *Discrete and Computational Geometry, Japanese Conference, JCDCG 2002 Revised Papers*, Volume LNCS 2866, pp. 159–171. Springer.

Frederickson, Greg N. (2005a). Hinged dissections: Swingers, twisters, flappers. In M. D. B. Cipra, E.D. Demaine and T. Rodgers (Eds.), *A Tribute to a Mathemagician*, pp. 185–195. Wellesley, Massachusetts: A K Peters, Ltd.

Frederickson, Greg N. (2005b). The manifold beauty of piano-hinged dissections. In R. Sarhangi and R. V. Moody (Eds.), *Renaissance Banff, Mathematics, Music, Art, Culture*, pp. 1–8.

Freese, Ernest Irving (1929a). The geometry of architectural drawing. Series of 22 articles in *Pencil Points*, 1929–1932. Fourteen more installments were written but never published.

Freese, Ernest Irving (1929b). Perspective projection. Series of five articles in *Pencil Points*.

Freese, Ernest Irving (1930a). Autobiography of Ernest Irving Freese. *Pencil Points* 11, 224.

Freese, Ernest Irving (1930b). Lines akimbo. *Pencil Points* 11, 711–720.

Freese, Ernest Irving (1939). Decoding the codes. Series of 14 articles in *Southwest Builder and Contractor*, 1939–1940.

Freese, Ernest Irving (1957a). Unpublished dissections.

Freese, Ernest Irving (1957b). *Geometric transformations. A graphic record of explorations and discoveries in the diversional domain of Dissective Geometry. Comprising 200 plates of expository examples.* Unpublished.

Friedlander, John B. and John B. Wilker (1980). A budget of boxes. *Mathematics Magazine* 53(5), 282–6.

Fujimoto, Shuzo (1982). *Suro Origami Asobi e no Shotai (Invitation to Creative Origami Play)*. Japan: Asahi Cultural Center.

G.-Mikusiński, Jan (1946). Sur quelques propriêtés du triangle. *Annales Universitatis Mariae Curie-Skłodowska Lublin-Polonia Sectio A* 1(2), 45–50.

Gardner, Martin (1983). *Wheels, Life and Other Mathematical Amusements*. New York: W. H. Freeman.

Golomb, Solomon W. (1994). *Polyominoes: Puzzles, Patterns, Problems, and Packings* (2nd ed.). Princeton: Princeton University Press.

Green, Chris (1998). *Triangular Philatelics: A Guide for Beginning and Advanced Collectors*. Iola, WI: Krause Publications.

Hanumanta Rau, B. (1888). *First Lessons in Geometry* (2nd ed.). Madras: S.P.C.K. Press. (S.P.C.K. is the Society for Promoting Christian Knowledge.) See page 127.

Harsdörffer, Georg Philipp (1657). *Vollständiges und von neuen vermehrtes Trincir-Buch.* Nürnberg: Verlag Paulus Fürsten.

Henderson, David W. (1996). *Experiencing Geometry: On Plane and Sphere.* Upper Saddle River, New Jersey: Prentice Hall.

Hoffmann, J. V. C. (1900). Ein anschaulicher Beweis des Satzes von der Winkelsumme des Dreiecks. *Zeitschrift für mathematischen und naturwissenschaftlichen Unterricht* 31, 263–264.

Hotchkiss, Philip K. (2002). It's perfectly rational. *College Mathematics Journal* 33(2), 113–117.

Houdini (1922). *Houdini's Paper Magic.* New York: E.P. Dutton.

Huffman, David A. (1976). Curvature and creases: a primer on paper. *IEEE Transactions on Computers* C-25(10), 1010–1019.

Hull, Thomas (2003). Rigid origami. Webpage (http://www.merrimack .edu/~thull/rigid/rigid.html).

Kelland, Philip (1855). On superposition. *Transactions of the Royal Society of Edinburgh* 21, 271–273 and plate V.

Lang, Robert J. (2003). *Origami Design Secrets: Mathematical Methods for an Ancient Art.* Natick, Massachusetts: A K Peters Ltd.

Langman, Harry (1950). Squaring the double-cross. *Scripta Mathematica* 16(4), 271.

Langman, Harry (1962). *Play Mathematics.* New York: Hafner.

Lemon, Don (1890). *The Illustrated Book of Puzzles.* London: Saxon.

Leonardo da Vinci (*Codex A*). *Il Codice Atlantico: della Biblioteca Ambrosiana di Milano.* Firenze: Giunti-Barbera. Transcription by Augusto Marinoni of the Codex Atlanticus, 1975.

Lindgren, H. (1951). Geometric dissections. *Australian Mathematics Teacher* 7, 7–10.

Lindgren, Harry (1952). Letter of February 28 to James Travers.

Lindgren, H. (1961). Going one better in geometric dissections. *Mathematical Gazette* 45, 94–97.

Lindgren, H. (1964a). Dissections for schools. *Australian Mathematics Teacher* 20(3), 52–54. The figures are in the supplement to the issue, pages i and ii.

Lindgren, Harry (1964b). *Geometric Dissections.* Princeton, New Jersey: D. Van Nostrand Company.

Lord, Nick (1990). The folding box problem. *Mathematical Gazette* 74(470), 361–365.

Loyd, Sam (*Eagle*). Loyd's Puzzles. Puzzle column in Sunday edition of *Brooklyn Daily Eagle* from March 22, 1896 to April 25, 1897. (1896a): April 12, p. 20.

Loyd, Sam (*Home*). Sam Loyd's Own Puzzle Page. Monthly puzzle column in *Woman's Home Companion*, 1903–1911. (1908a): Nov., p. 51.

Loyd, Sam (*Inquirer*). Mental Gymnastics. Puzzle column in Sunday edition of *Philadelphia Inquirer*, October 23, 1898–1901. (1900a): April 22; (1901a): August 11.

Loyd, Sam (*Press*). Sam Loyd's Puzzles. Puzzle column in Sunday edition of *Philadelphia Press*, Feb. 23–June 29, 1902. (1902a): March 16.

Loyd, Sam (*Tit-bits*). Weekly puzzle column in *Tit-Bits*, starting in Oct. 3, 1896 and continuing into 1897. Dudeney, under the pseudonym of "Sphinx," wrote commentary and handled the awarding of prize money. Later in 1897 Dudeney took over the column. (1897a): April 3, p. 3; (1897b): April 24, p. 59.

Lurker, Ernst (1984). Heart pill. 7 inch tall model in nickel-plated aluminum, limited edition of 80 produced by Bayer, in Germany.

MacMahon, Percy A. (1922). Pythagoras's theorem as a repeating pattern. *Nature* 109, 479.

Mahlo, Paul (1908). *Topologische Untersuchungen über Zerlegung in ebene und sphaerische Polygone.* Halle, Germany: C. A. Kaemmerer. Ph.D. dissertation for the Vereinigte Friedrichs-Universität in Halle-Wittenberg. See pp. 13, 14 and Fig. 7.

Miura, Koryo, Masamori Sakamaki, and K. Suzuki (1980). A novel design of folded map. In *Abstracts of Papers: the 10th International Conference of the International Cartographic Association*, Tokyo, pp. 216–217. Poster session.

Mott-Smith, Geoffrey (1946). *Mathematical Puzzles for Beginners and Enthusiasts.* Philadelphia: Blakiston Co. Reprinted by Dover Publications, New York, 1954.

Odani, Kenzi (2000). Maximal volume of curved folding boxes. *Mathematical Gazette* 84(499), 110–113.

Olson, Alton T. (1975). *Mathematics through Paper Folding*. Reston, Virginia: National Council of Teachers of Mathematics.

Ozanam, Jacques (1778). *Récréations Mathématiques et Physiques*. Paris: Claude Antoine Jombert, fils. See figures 123–126 and pages 297–302. This is material added by Jean Montucla, who is listed as a reviser under the pseudonym of M. de Chanla. See also pages 127–129 of *Recreations in Mathematics and Natural Philosophy*, by Jacques Ozanam, London, Thomas Tegg, 1840, translated from Montucla's edition by Charles Hutton, with additions.

Paterson, David (1989). T-dissections of hexagons and triangles. *Journal of Recreational Mathematics* 21(4), 278–291.

Paterson, David (1995). Dissections of squares. Unpublished.

Perigal, Henry (1891). *Graphic Demonstrations of Geometric Problems*. London: Bowles & Sons. On cover: "Geometric Dissections and Transpositions," Association for the Improvement of Geometrical Teaching. (The association was later renamed The Mathematical Association.).

Petroski, Henry (2005). Industrial origami. *American Scientist* 93(1), 12–16.

Pirich, Donna Marie (1996). A new look at the classic box problem. *PRIMUS* 6(1), 35–48.

Pook, Les (2003). *Flexagons Inside Out*. Cambridge: Cambridge University Press.

Reed, Neville (1992). A curved folding box. *Mathematical Gazette* 76(476), 275–277.

Reid, Robert (1987). Disecciones geometricas. *Umbral* (2), 59–65. Published in Lima, Peru, by Asociacion Civil Antares. (The author's name as listed in the article is Robert Reid Dalmau, conforming to Spanish custom, but is listed here in the form that Robert prefers.)

Rosenbaum, Joseph (1947). Problem E721: A dodecagon dissection puzzle. *American Mathematical Monthly* 54, 44.

Ross, W. W. (1891). *Mensuration Taught Objectively, with Lessons on Form*. Fremont, Ohio. Manual for the use of the author's dissected surface forms and geometrical solids.

Rubik, Erno (1983). Toy with turnable elements for forming geometric shapes. U.S. Patent 4,392,323. Filed 1981; filed for a Hungarian patent in 1980.

Sayili, Aydin (1958). Sābit ibn kurranin Pitagor teoremini temini. *Türk Tarih Kurumu. Bulleten* 22, 527–549.

Sayili, Aydin (1960). Thâbit ibn Qurra's generalization of the Pythagorean theorem. *Isis* 51, 35–37.

Slocum, Jerry and Jack Botermans (1986). *Puzzles Old & New*. Plenary Publications International (Europe) bv, De Meern, the Netherlands, and ADM International bv, Amsterdam, the Netherlands. See pages 148–149.

St. André, Richard (1983). The box problem. *Mathematics Teacher* 76(2), 108–109.

Stanley, Dick (2001). The box problem. Draft.

Steinhaus, Hugo (1950). *Mathematical Snapshots*. New York: Oxford University Press.

Stevens, Kenneth V. (1994). Folding puzzle using triangular pieces. U.S. Patent 5,299,804. Filed 1993.

Sturm, Johann Cristophorus (1700). *Mathesis Enumerata: or, the Elements of the Mathematicks*. London: Robert Knaplock. Translation (by J. Rogers?) of 1695 work *Mathesis Enumerata*. See pp. 20–21 and Fig. 29.

Sundara Row, T. (1893). *Geometrical Exercises in Paper Folding*. Madras: Addison. See page 68.

Taylor, H. M. (1905). On some geometrical dissections. *Messenger of Mathematics* 35, 81–101.

Theobald, Gavin (2004). Geometric dissections. Collection of webpages (http://home.btconnect.com/GavinTheobald/Index.html).

Todhunter, I. (1852). *A Treatise on the Differential Calculus, and the Elements of the Integral Calculus, with Numerous Examples* (first ed.). London: Macmillan. See problem 28 on page 193.

Torbijn, Pieter (2001, October). Hinging from pentomino to square. *Cubism For Fun* (56), 17–19.

Valens, Evans G. (1964). *The Number of Things*. New York: Dutton.

van Schooten, Frans (1659). Geometriam renati des cartes commentarii. In *Geometria, a Renato Des Cartes anno 1637 gallicè edita*. Amstelædami: Ludovicum & Danielem Elzevirios. See page 263.

Varignon, Pierre (1731). *Elemens de mathematique de Monsieur Varignon*. Paris: Pierre-Michel Brunet, fils. See "Partie II. Elemens de Geometrie", Corollaire IV, pages 64–65, and Table 8, Fig. 11.

Varsady, Alfred (1989). Some new dissections. *Journal of Recreational Mathematics* 21(3), 203–209.

Index of Dissections

Dissections are ordered by the following conventions:

1. The geometric figures are ordered with those represented by letters (in alphabetical order) following those represented by $\{p\}$ or $\{p/q\}$ (in lexicographic order on (p, q)).

2. A dissection is listed under the figure it involves that comes latest in the list.

3. For a figure such as $\{r\}$, dissections of it involving $\{p\}$ or $\{p/q\}$ with $p < r$ come first. Within dissections only involving $\{r\}$, special relationships come first (ordered lexicographically), then general relationships (ordered lexicographically), then a $\{r\}$s to b $\{r\}$s (ordered lexicographically on (a, b)).

Index

Printed in the United States
by Baker & Taylor Publisher Services